AFRICA COUNTS

Number and Pattern in African Culture

CLAUDIA ZASLAVSKY

with an introduction by John Henrik Clarke,
President of African Heritage Studies Association

Lawrence Hill Books

DEDICATION

This book is dedicated to those Africans who are now engaged in the formidable task of reclaiming their heritage.

Paperback edition, May 1979
Published by Lawrence Hill Books
Brooklyn, New York 11238
Lawrence Hill Books is an imprint of
Chicago Review Press, Incorporated
Chicago, Illinois 60610

ISBN 1-55652-075-1, previously
       0-88208-104-7

Printed in the United States of America.

Reprint by permission of Prindle, Weber & Schmidt,
Boston, Massachusetts   02116

CONTENTS

INTRODUCTION

This is a pioneer work in an area of African history and culture that is virgin territory for the researcher. Except for the unpublished manuscript of Frank E. Chapman, Jr., the lectures of Dr. Lonnie Shabazz, formerly head of the Department of Mathematics at Atlanta University, and several other sources listed in the bibliography of this book, I know of no extensive work on African mathematics. There is also a tendency on the part of most Western scholars to deny the existence of mathematical methods that are distinctly African. Therefore, Mrs. Zaslavsky's "preliminary survey of a vast field awaiting investigation" will put some old arguments to rest and create some new ones.

It is impossible to write a book of this nature without some reference to the misconceptions about Africa in general. Fortunately, the book is being published at a time when information about African history and culture is increasing and many of the new and important books are coming from the pens of Black writers, in this country and in Africa. It will soon be difficult to ignore the impact of Africans on world history.

Now that the field of mathematics, as it relates to Africans, has been opened up for investigation by Mrs. Zaslavsky's book, it will be difficult to stop the much broader investigation into the role of Africans in world history and culture. This book is important both for what it says and for how it implies what still needs to be said. It is clear that the old concepts about Africa and the African people will die hard, *but they will die.*

John Henrik Clarke

Associate Professor
Department of Black and
    Puerto Rican Studies,
    Hunter College
President of the African
    Heritage Studies Association

PREFACE

This book is offered as a preliminary survey of a vast field awaiting investigation. To my knowledge, no analysis of African culture from the mathematical point of view has ever been published. True, there are numerous studies of numeration systems and such specialized subjects as games having a mathematical content, all of which I have found valuable.

It was with trepidation that I approached African-born scholars to gather material for this book. Would they consider me, a white American, presumptuous in writing about *their* culture? I was careful to emphasize that I was directing my appeal to American audiences. The reaction to my requests was almost uniformly enthusiastic. African scholars are so overburdened by the enormous task of reconstructing their history, recording their traditions and setting the record straight, that they welcome assistance from sympathetic researchers of any origin. Hopefully, as investigators attain a deeper knowledge of Africa's past, its contributions to mathematics will be brought to the attention of the world.

The inadequacy of easily accessible material is the chief difficulty in writing a book such as this. One must search the literature of many disciplines—history, economics, ethnology, anthropology, archaeology, linguistics, art and oral tradition—and still be dissatisfied. African literature —novels, poems, essays—sometimes tells more about mathematics in the context of the lives of the people than do articles in learned journals. To correct the deficiency, ideally, representatives of each ethnic group should write down for posterity the mathematical ways of their forefathers. Or a researcher should live with each group, become part of the community by learning the language and adopting the customs, and hopefully begin to understand how mathematics influences the lives of the people. Both methods are being used, and we hope that there will be ample records before the old cultures give way to modern technology, and the local languages yield to a national language, possibly English or French!

A technique I used in my research was to summarize the available literature, mostly of European origin, and then check the material with people of the ethnic background I was investigating, usually young people of the faculty or student body of schools and universities. I would read from my notes: "The numbers three, five and seven are considered unlucky." "Never heard of such a thing," might be the reply. In such cases I was unable to reach a satisfactory conclusion as to whether the European investigator was ill-informed, or the beliefs had changed since the early report, or possibly both conditions were true. One of my informants, a middle-aged man, promised that when he visited his homeland he would question his father, a person well versed in magic. Soon afterward the younger man was called to the diplomatic service of his country, and I never received the information. Many of the young people said they would have to check with their parents, or that a few old people still had such a custom, but that most people "didn't bother with it." Nevertheless, many customs noted in this book are very much alive and flourishing. I have followed the procedure of writing in the present tense except where I am sure that only the past tense is correct.

Another difficulty is the great number of African languages. The American linguist J. H. Greenberg lists the names of seven hundred thirty languages, but he maintains there are well over a thousand. He has classified them into several families of which the largest is Bantu, spoken by most of the inhabitants of the southern half of Africa.

A related problem is that of the names of ethnic groups, variations in the names, and variations in their spellings. A people may call themselves by a certain name, but may have different appellations in the languages of their neighbors, the Arabs, and Europeans of various nationalities. Take this example: according to H. A. Johnston (*The Fulani Empire of Sokoto*, Oxford, 1967) Fulani is the name applied by the English to a certain people inhabiting the Sudanic area; however, they call themselves the Fulbe, and they are known as the Peul to the French, Fula to the Bambara, Felaata to the Kanuri, and Fillani to their Hausa neighbors in northern Nigeria. The people of Mali were variously known as Mandingo, Mandinke, Malinke and Mande.

When I interviewed Dr. Kipkorir of the University of Nairobi, he handed me a chart entitled "Kalenjin Speaking Peoples," published by the Committee on Historical Terminology of the Historical Association of Kenya. The headings are: A. Common Reference in Published Works; B. Name by which group knows itself; C. Alternative names applied to group; D. Spelling in accordance with local pronunciation. Everywhere studies have been undertaken to correct and standardize nomenclature.

I had read an article on the numeration system of the Dabida people, living in the Taita Hills in Kenya, but no ethnographic reference book listed their name. While in Kenya I was introduced to a young college teacher, a native of the Taita Hills. "Have you heard of the Dabida people?" I asked Mr. Irina. He replied with a smile. "My people call themselves 'Dabida.' Usually they are identified as 'Taita' in the literature."

As for the number words themselves, sources of information often disagree on the names of the numbers, the meaning of the names, and even their spellings. Frequently the words have changed over the years. In researching the Igbo (frequently spelled "Ibo") numeration, I used several works published during the past sixty years. I found many differences in spelling, and several usages peculiar to particular local dialects. Mr. A. A. Dike, an Igbo sociologist, was most helpful in pointing out the closest to "standard" Igbo numeration.

I have tried to use names and spellings most acceptable to Africans: Igbo, Maasai, Asante, rather than Ibo, Masai, Ashanti; Lake Nyanza for Lake Victoria, Namibia for South West Africa, Zimbabwe for Rhodesia. I have used the familiar terms "Bushmen" and "Pygmies," in addition to identifying these peoples by group, for the benefit of the non-African reader. The names of Bantu-speaking peoples are given without the prefix: Tutsi and Kamba, instead of the plural forms Batutsi (or Watutsi in Swahili) and Akamba. Properly, one would say that the people known as the Baganda (singular, Muganda) live in Buganda and Luganda. All diacritical marks have been omitted when writing words in African languages.

I apologize for any errors that I have committed in dealing with concepts, as well as in terminology, spelling, or usage.

SECTION 1
THE BACKGROUND

Was there mathematics in Africa? On the basis of the references to Africa in books about the history of mathematics and the history of number, one would conclude that Africans barely knew how to count. In the opening section we shall summarize the content of this book as contrasted with the treatment of African numeration by American and European authors. A brief history of Africa, home of man's most remote ancestors, includes an account of ancient Egyptian mathematics as well as a look at several centuries-old African societies.

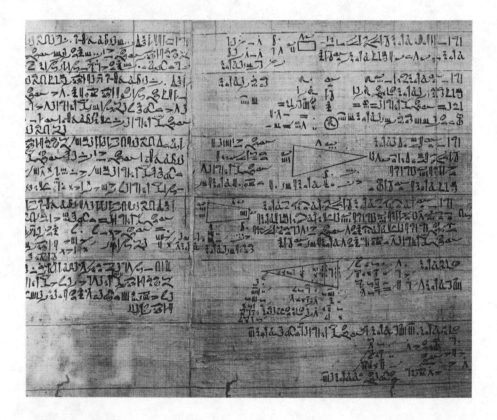

Figure 1–1   Extract from the Rhind papyrus, written by the Egyptian scribe Ahmes about 1650 B.C. This document and the Moscow papyrus are our chief sources of knowledge of ancient Egyptian mathematics. This text contains 85 problems of a practical nature. British Museum.

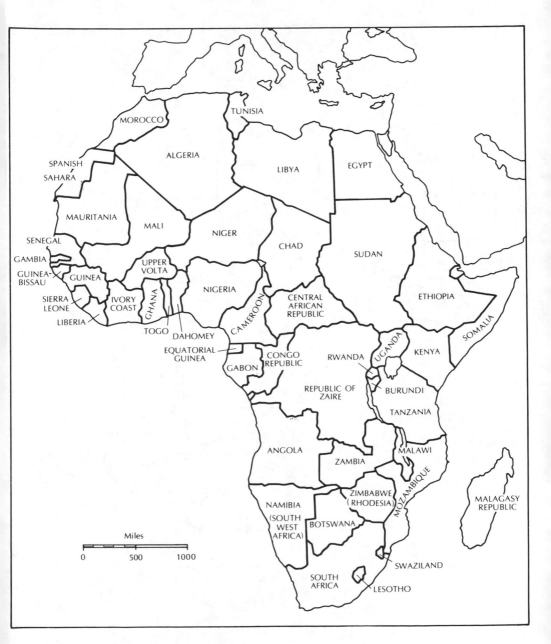

Figure 1-2 Africa Today

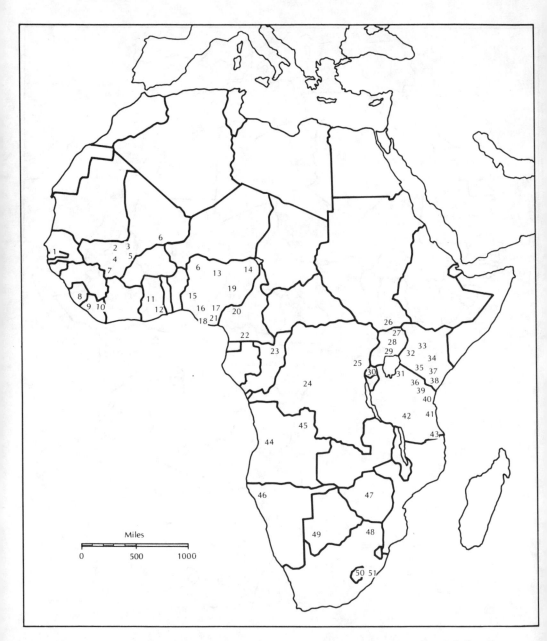

Figure 1–3   Peoples Discussed in the Text. (See complete listing on following page.)

Peoples referred to in the text.

11. Akan (includes Asante)	24. Kuba (Shongo, Bushongo)
36. Arusha Maasai	23. Kwele
11. Asante (see Akan)	32. Luo
1. Balante	35. Maasai
4. Bambara	43. Makonde
16. Bini (Edo)	7. Malinke (Mandingo)
49. Bushmen (San)	44. Mbundu
24. Bushongo (Shongo, Kuba)	25. Mbuti (Pygmy)
39. Chagga	8. Mende
5. Dogon	33. Nandi (see Kalenjin)
16. Edo (Bini)	29. Nkole (Hima)
21. Efik	25. Pygmy (Mbuti)
20. Ekoi	30. Rwandese (includes Tutsi,
12. Ewe	Hutu, and Twa)
6. Fulani	49. San (Bushmen)
28. Ganda	40. Shambaa
13. Hausa	47. Shona
42. Hehe	24. Shongo (Bushongo, Kuba)
46. Herero	3. Songhai
29. Hima (Nkole)	2. Soninke
26. Huku	50. Sotho
17. Igbo	41. Swahili
18. Ijo	38. Taita
45. Jokwe	19. Tiv
33. Kalenjin (includes Kipsigis and Nandi)	30. Tutsi (included in Rwandese)
37. Kamba	9. Vai
14. Kanuri	48. Venda
27. Karimojong	22. Yaounde (Beti)
34. Kikuyu	15. Yoruba
33. Kipsigis (see Kalenjin)	31. Ziba
10. Kpelle	51. Zulu

CHAPTER 1
AFRICAN MATHEMATICS?

Before the time of Newton, mathematics and science had little impact on the daily lives of people. Since then mathematics and technology have worked hand in hand to bring about radical innovations in our life style. In an "advanced" country like the United States, the greatest portion of the federal budget is now spent on products whose existence depends on higher mathematics—military hardware and equipment for space travel. People are constantly being urged to purchase more cars, more electronic devices, more material goods manufactured by precisely designed machinery. The computer has taken over many jobs of office workers, technicians, factory operators and teachers. Technology, made possible by mathematics, has become the new god and the possession of its products the goal of life. Yet, the ways of this new god remain as mysterious to the uninitiated citizen of the modern world as were those of the ancient deities.

To the average person, mathematics is a subject that deals with numbers in counting and the computations of elementary arithmetic. I have had my students tell me, "I could do the proof all right, but I made a mistake in the math. I added two and three, and got six!" So, when I tell people that I am doing research in African mathematics, they often ask: "How *do* Africans count—like us?" I posed this question to two young colleagues, one from Kenya and the other from Tanzania. "How do you count in your own language?" They looked at me in surprise, and replied, "Just as you do here in the United States!" Nowadays school children in most of the world write Hindu numerals (popularly called "Arabic")—truly universal symbols.

The Sociomathematics of Africa

For many mathematicians, mathematics as a discipline originated in ancient Greece with the formulation of logical systems based on definitions, postulates and formally proved theorems. A more flexible definition—the British science writer, Lancelot Hogben calls it a "provisional formula"—is offered in his *Mathematics in the Making* (page 9): "Mathematics is the technique of *discovering* and *conveying* in the most *economical* possible way *useful* rules of reliable reasoning about *calculation, measurement,* and *shape.*"

The last part of Hogben's provisional formula describes the subject matter of this book; it deals with "calculation, measurement, and shape." It would be difficult to discuss "discovery", since few records exist. "Conveying" was accomplished chiefly by word of mouth and by example, from one generation to the next. "Useful" is the key word; I shall deal with mathematics entirely from the point of view of its applications, and not as an abstract body of thought.

It is true that Africans, to our present knowledge, have only recently begun to participate in the development of pure mathematics; possible reasons will be suggested in the concluding chapter. But mathematics is in evidence in many aspects of African life. This book is concerned with the "sociomathematics" of Africa—the applications of mathematics in the lives of African people, and, conversely, the influence that African institutions had upon the evolution of their mathematics.

Geographically, this study is confined to that part of Africa which lies south of the Sahara, an area comparatively isolated from the Mediterranean, Europe, and Asia. We are concerned with a vast region extending nearly four thousand miles from north to south and over four thousand miles from east to west. Within these boundaries exist a variety of cultures and a corresponding diversity of mathematical developments.

The development of a number system depends upon need. In a small, self-contained economy in which all or most of the necessities of life are produced within the community—typical of large sections of Africa—there is little need for an extensive reckoning system. The names of numbers are frequently connected with the objects to be counted, just as we have special names for certain sets—flock, herd, brace, etc., dating back from a pastoral or agricultural society. Gesture-counting is especially necessary in the market place, where people speaking various languages gather to exchange goods. It may be customary to use beads, shells, nuts, or pebbles (the word calculate is derived from *calculus*, Latin for pebble) as media of exchange or counting materials, and to arrange them in sets, thus giving rise to special words.

Systems of numeration range from the few number words of some San peoples, who have been pushed into the least hospitable areas of the continent by advancing black and white populations, to the extensive numerical vocabulary of nations having a history of centuries of commerce. A characteristic of African counting is a standardized system of gestures to accompany, or even replace, the number words. The gesture languages show as much variety as the spoken ones.

Mystical beliefs about numbers are many and varied. Certain numbers are deemed to have special significance, favorable or unfavorable. In Chinua Achebe's novel *Arrow of God* the medicine-man counted the cowrie shells "carefully on the ground as a woman would before she bought or sold in the market in groups of six. There were four groups and he nodded his head" (p. 147). The Igbo were unique in counting cowries in units of six, and the number four has particular significance for them.

In many societies it was the use of cowrie shell currency which created both the need for number names in the higher denominations and special categories of numbers. The demands of commerce also dictated the degree of standardization of weights and measures. The villager with the longest arm sometimes set the standard for measuring cloth! Record keeping varied from knotted strings and notched sticks to the ritual of the annual census, carried out by indirect methods that circumvented the taboo on counting living creatures.

The ability to observe and reproduce patterns, both numerical and geometrical, is of great importance in Africa, where most societies have been nonliterate until recently. The Bushman in the Kalahari Desert walks miles to dig up a watery root whose location he had noted several months previously—and with no man-made markers to guide him! Cattle-herding folk have in their vocabularies dozens of words to describe their livestock on the basis of hide markings and dozens more to differentiate cattle by the shape of the horns. Each pattern in weaving, in wood carving, in cloth dying, has a special meaning. Numerical patterns are evident in games, from versions of tic-tac-toe to the universal African board game in which seeds or pebbles are transferred and captured. The scholars of Muslim West Africa associated astrology and numerology with arrays of numbers called magic squares.

African art is varied and complex; geometric patterns and symmetries appear in the beautifully carved and painted masks associated with religious practices and in the decoration of such common household objects as gourds and baskets. Patterned mats and carved wooden doors embellish the home in some cultures; in others a round thatched cottage provides temporary shelter until the herds of cattle move on to greener pastures. Geometry points to a social distinction in the Kikuyu village of James Ngugi's *The River Between*, where the rectangular house of the one converted Christian is conspicious among the round homes of his neighbors.

The general discussion of African numerical and geometric concepts in this book will be followed by studies in detail of southwest Nigeria and East Africa. These will enable the reader to observe closely the development of mathematical concepts in two areas having quite different historical and cultural backgrounds. The Yoruba people and the related people of Benin, in Nigeria, have been urbanized farmers and traders for centuries and have a complex numeration system noted by many investigators. East African society is varied—pastoral and agricultural, stateless and highly centralized—with cattle the mainstay of the economy.

The book concludes with a discussion of the disruptions caused by the slave trade and colonialism during the past five centuries, and their disastrous effect upon the potential development of science and technology in Africa.

Africa's Place in Writings on the History of Mathematics

Africans have made significant contributions to the development of count-ing and numbers and deserve a place in studies dealing with this subject. But what do we find? Even recently published books suffer from inade-quacy of material, exhibit an outdated point of view, and repeat incorrect information.

Until the nineteenth century, Europeans had few contacts with Africa other than commerical dealings and the slave trade. In the late nine-teenth century, stimulated in part by missionaries and commerical adven-turers, the European countries engaged in a scramble for African territory. The continent was divided among the powers, who sent in administrators to set up colonial governments. They were soon joined by anthropologists. In a short time numerous studies became available of the many ethnic groups, explaining their customs, religion, dress, language, art—every facet of life. It was necessary to describe their numeration systems; how else could the colonial administrations employ Africans for wages and levy taxes upon them? Much valuable descriptive material became avail-able to European and American readers.

In Great Britain there arose a school of anthropologists, led by E. B. Tylor, having a point of view based on their interpretation of the new doctrine of evolution. Their thesis was this: man evolved from a primitive to an advanced state during the course of many millennia. The white man had arrived at the highest level, in contrast to the "primitive savages" of "Darkest Africa," who were still in the very early stages of evolution. Tylor's *Primitive Culture* became the leading reference work for anthropologists, ethnologists, and even for writers of the history of mathematics. Using Tylor as his source, the mathematical historian Florian Cajori wrote in 1896 (page 3):

> Of the notations based on human anatomy, the quinary and vigesimal
> systems are frequent among the lower races, while the higher nations
> have usually avoided the one as too scanty and the other as too cum-
> brous, preferring the intermediate decimal system.

The quinary and vigesimal systems are common in Africa!

L. L. Conant's *The Number Concept*, published in 1896, is one of the few books in English which discuss the numeration systems of many African peoples. However, Conant's point of view is completely colored by the pre-vailing attitude toward Africans as "primitive savages"; they were deemed hardly human. He dismisses the amazingly complex numeration system of the Yoruba people of southwest Nigeria with these words (page 32):

> Nor on the other hand, is the development of a numeral system an in-
> fallible index of mental power, or of any real approach toward civiliza-

tion. A continued use of the trading and bargaining faculties must and does result in a familiarity with numbers sufficient to enable savages to perform unexpected feats in reckoning. Among some of the West African tribes this has actually been found to be the case; and among the Yorubas of Abeokuta the extraordinary saying, "You may seem very clever, but you can't tell nine times nine," shows how surprisingly this faculty has been developed, considering the general level of savagery in which the tribe lived.

Conant sees the occurrence of numbers up to a million among South African people as "remarkable exceptions" to the "law" that "the growth of the number sense keeps pace with the growth of the intelligence in other respects" (p. 33). Such was the extent of the prejudice against dark-skinned peoples that he turned all the principles of logic upside down in contradiction of the scientific method. One would expect that, when confronted with evidence that refutes his "law," the scientist should begin to doubt its validity or else apply his "law" to draw the logical conclusion about the high level of intelligence of his subjects.

Lévy-Bruhl's *How Natives Think*, originally published in French in 1910, contains excellent criticism of the Tylor school of anthropology, and of the data-gathering techniques upon which they relied (page 20):

They observed what seem to them most noteworthy and singular, the things that piqued their curiosity; they described these more or less happily. . . . Moreover, they did not hesitate to interpret phenomena at the time they described them; the very idea of hesitation would have seemed quite unnecessary. How could they suspect that most of their interpretations were simply misapprehensions, and the "primitives" and "savages" nearly always conceal with jealous care all that is most important and most sacred in their institutions and beliefs?

These false interpretations were copied from one book to another, and translated from one European language to another. Lévy-Bruhl's own work was marred by his distinction between the "pre-logical" or "mystical" mentality of the "lower societies" (*les sociétés inférieures*), and the "logical" mental activity of civilized peoples.

This "scientific" attitude is analyzed in an unpublished manuscript, *Science and Africa*, by Frank E. Chapman, Jr., a brilliant young black man now serving life imprisonment. (See the Appendix for his autobiography.)

There is so much talk about a "primitive type of mind" and an "advanced type of mind." These conceptualists rarely pause to consider what it is in fact that they are talking about; they never stop to consider that this whole business of "mind types" is merely a collocation of convenient verbalizations; indeed, the scientist's behavior is very similar to the "primitive" he is talking about when he verbalizes about "mind types." This stifling "mind" concept cripples scientific analysis, and only when greater emphasis is put on psychological behavior patterns (in a given social context) will it be understood once and for all ... that the difference in psychological make-up is due, more or less, to the differences in social conditions, which have nothing whatsoever to do with "mind types."

Karl Menninger's *Number Words and Number Symbols*, published in 1969 in an English translation of the 1958 revised German edition, deals

Figure 1–4 Maasai married women wearing coiled neck rings. Peabody Museum, Cambridge, Mass.

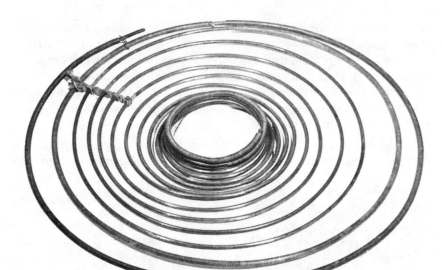

Figure 1–5   Coiled iron necklace: a neck ornament consisting of one long metal wire that coils away from the neck in such a way that it extends over the shoulders. The inner two rings and the outer two are worked with thin wire (brass), and the outer six rings have been bound together with leather binding. National Museum of Tanzania.

extensively with the development of numeration systems throughout the world. On African systems there is very little: a few number words of two unspecified languages, Francis Galton's story of the difficulty of purchasing sheep from the "primitive" Damara people, the repetitive counting of the Bushmen, a "Masai girl wearing annual rings, which show her to be 23 years old" (page 33). This caption accompanies a photograph of a young woman wearing a spiral metal collar. The text states: "Young unmarried girls of the Masai, a warlike tribe of herdsmen living on the slopes of Mt. Kilimanjaro, each year add one heavy brass ring around their necks, so that their precise age can be known from this extraordinary necklace of annual rings." Hollis, in his book *The Masai*, published in 1905, has the identical photo, as well as a picture of a married woman wearing a similar necklace with only three coils. He makes a clear distinction between the style of ornament worn by uncircumcised girls and those worn by women. However, I was told by Mr. Ole Kantai, a Maasai instructor at the University of Nairobi, that a coiled necklace is worn only by a married woman as a symbol of her husband's affection. I also described the Menninger interpretation to Mr. Onesmo ole Moiyoi, a young Maasai now studying at an American medical college. He wrote:

The number of rings in a Maasai girl's necklace has no bearing what-ever on her age. Such a statement is, I think, a reflection of the lack of understanding regarding the mechanics of making iron necklaces. The caption implicitly points out the necessity of making a necklace once every year for each girl that wears one. This is not done.

What impression of African numeration systems does one gain from reading Menninger's book? The only counting system he identifies by name is that of the Bushmen, perhaps the oldest people in Africa. "The Bushmen of South Africa have only the two number words a, 'one', and oa, 'two', and they say 'four' as oa oa and 'five' as oa oa a. . . This is not a gradation based on two but merely an ordering, a much earlier phase, especially if it occurs in speech" (page 64). Yet even with regard to this elementary type of counting there is more to be said. The Viennese anthro-pologist, Marianne Schmidl, who also used secondary sources, but with greater care and judgment, writes of several different Bushman languages. In one, there are words for one and two, and then a third word meaning "many," while in another there are three number words, then a fourth meaning "many." In a third Bushman language the number words refer to the number of fingers, hands, and even feet required to represent the number; for example, fifteen is "two hands, one foot."

When I was interviewing faculty members at the University of Nairobi to obtain material for this book, I introduced my topic by saying that I was planning to write for American readers, whose impression of African numeration, based on the mathematics history books available to them, was that Africans could only count "one, two, many." One of the instructors immediately picked up the phone. "I have an American lady here, who says that Americans think Africans count 'one, two, many.' Of course, we know it is the Australian aborigines who count 'one two, many'!"

Today, less than three generations after Conant and Cajori wrote about "primitive" minds, Africans are computerizing their government's operations and filling the chairs of mathematics in their universities. An-thropologists and other researchers in African culture have long since revised their attitudes. It is time for the mathematical historians to catch up with them.

The Contribution of Marianne Schmidl

"The standard authority on the number systems in Africa" was David Eugene Smith's description of Marianne Schmidl in his *History of Mathe-matics*. Her work, "Zahl und Zählen in Afrika," published in German in

1915, has never been translated into English, although writers in the English language have borrowed heavily from it, not always with full acknowledgment of their source. I have found no later work equal to it. Dr. Schmidl, knowing that all of her two hundred sixty references were the work of European researchers in Africa—anthropologists, administrators, missionaries—began her paper on a note of caution (p. 165):

A number of works are available on primitive counting and the development of number concepts, but specialized comparative research is scarce. The following paper attempts to survey conditions in Africa, a region about which no such work is available.

I have not found a completely satisfactory solution to the task I assigned myself for the following reasons. Above all, there is a lack of material so that the picture of African counting that we can give today is incomplete. Also the different types of numerical expressions occur infrequently and are incompletely described in the literature; therefore it would not be correct to arrive at definite conclusions on the basis of these scattered notes. New research should be conducted with the native peoples themselves in their own homelands, or at least in the linguistic seminars, as in Hamburg and Berlin, where native-born Africans are available. . . .

She sought to correct the errors of previous writers, who assumed that there was a parallel development throughout the world, and then attempted to prove their thesis by offering examples from different regions. Schmidl maintained that development is unequal, and that the historical and linguistic factors must be examined in connection with each ethnographic region. She concluded (page 196):

True primitiveness in relation to counting cannot be found. Even where the expression for 3 or 4 means "many," as for example among the Bushmen, it is not true that all counting ends there. They continue counting with the assistance of the words for two and one, and when they do not use higher numbers, they assign a specific meaning to the word "many" by means of appropriate gestures.

One must be extremely cautious about accepting the accounts of the inability of "primitive" people to count in higher denominations. ("Primitive" must be taken here, as elsewhere, with a grain of salt!) The absence of counting words by no means indicates a lack of counting concepts, since the number concept and the designations for the numbers need not always coincide. One must consider the level of the economy, the practical need for arithmetic operations, etc.

I must mention another phenomenon: the reluctance of primitive people to count human beings, animals, or valuable possessions, for fear of incurring the envy of evil spirits.

Such views were incompatible with those of the Nazi conquerors of Austria. During the German occupation Dr. Schmidl was killed—a victim of Hitler's policy of extermination of the Jews. (See the Appendix for a biographical note.)

Differences in Mathematical Development

There is general agreement among anthropologists that in the evolution of counting, number was originally conceived in connection with the objects to be counted, and it was only at a much more advanced stage of development (like ours) that man conceived of number in the abstract.

My experience as a mathematics teacher has taught me that most people in our own society still think of numbers in concrete terms. The arrogant attitude toward Africans of Adolphus Mann, in his report to the Royal Anthropological Society in 1887, was more sweeping in its application than he intended. He followed a description of the complex numerical system of the Yoruba people with this remark: "Other objects are only counted in comparison with an equal number of cowries [the cowrie shell currency then in use], for a nation without literature and without a school knows nothing of abstract numbers" (page 61).

I have found that many people in modern society have difficulty in handling abstract numbers. Some years ago I was helping a bright secondary school student with her algebra. (She now holds a responsible position in publishing.) The problem required that she find the product of 3 and 25. She wrote

$$\frac{\begin{array}{r} 25 \\ 3 \end{array}}{}$$

and proceeded to multiply at length, as she had been taught in the elementary grades. I interrupted to ask, "Vicky, how much money is three quarters?" "Seventy-five cents," she replied without hesitation, but with an expression of surprise. I find that even in our highly mathematical culture the technique of referring to currency helps to clear up many difficulties in dealing with fractions, decimal points, percent, and other computations.

Of course there is a world of difference between the mathematical thinking of the hunter and gatherer in the Kalahari Desert, and the trader, African or American, who is accustomed to dealing in large sums of currency. What factors account for the difference? According to Raymond L. Wilder, in *Evolution of Mathematical Concepts* (page 4), "Mathematics is something that man himself creates, and the type of mathematics he works out is just as much a function of the cultural demands of the time as any of his other adaptive mechanisms." Today mathematics has developed to such an extent that even the most brilliant mathematician is incapable of knowing all the aspects of this field. This advanced level of mathematics cannot be attributed to the "superiority of the modern mind." "There is not one iota of anatomical or psychological evidence to indicate that there are any significant innate, biological or racial differences so far as mathematical or any other kind of human behavior is concerned," writes the anthropologist Leslie A. White (page 2353). He says further (page 2358):

> The whole population of a certain region is embraced by a type of culture. Each individual is born into a pre-existing organization of beliefs, tools, customs and institutions. These culture traits shape and mould each person's life, give it content and direction. Mathematics, is, of course, one of the streams in the total culture.

In every period of time, in every geographical location, people inherit from their ancestors their ways of life. The resources available to the society in the form of materials, ideas, and human relationships determine the changes that take place in their institutions, including their mathematics. Great minds have their part in mathematical discovery, but they can operate only within the cultural setting in which they exist.

In Africa, too, physical environment, inherited cultural resources, and the impact of external forces determined the nature and extent of mathematical development.

CHAPTER 2

HISTORICAL BACKGROUND

Africa in the Prehistoric World

Man apparently first emerged on the African continent. The scientific world was astounded when Drs. Mary and Louis Leakey announced the discovery in 1959 of the remains of *Homo habilis*, a tool-making hominid estimated by the potassium-argon dating technique to be 1,750,000 years old. The scene of the discovery, Olduvai Gorge in northern Tanzania, has since yielded much information about the development of man.

Later, two human skulls and stone tools were excavated in Kenya, east of Lake Rudolf; their age is estimated to be 2.6 million years. In February, 1971 came the announcement that a human type of jawbone from the same area was five and a half million years old! No part of the world has proved as fertile a field for the discovery of skeletal remains and stone tools. The most impressive to date is the finding at Fort Ternan, Kenya, of a man-like creature and animals estimated to be fourteen million years of age.

A spectacular discovery, reported early in 1970, was that of an ancient mine in an iron-ore mountain in Swaziland, in southeast Africa. Stone age mining tools were found, and samples of charcoal remaining from old fires were tested by the radio-carbon dating technique. The mine turned out to be 43,000 years old! The ancients appeared to have been seeking specularite, a valuable pigment and cosmetic.

Rock paintings, executed ten thousand years ago, point to an era when the Sahara was green and lush, inhabited by cattle and horses, and by human beings who hunted, rode in chariots, danced, worshipped, and loved. Other such paintings, some dating back thirty thousand years, have been found in South and East Africa, giving us clues to the lives of the people who lived in ancient times.

During the later stone age, hunting and fishing societies developed in the Nile Valley, in West Africa, and in East Africa. From the mathematical point of view the most interesting find is a carved bone discovered at the fishing site of Ishango on Lake Edward, in Zaire (Democratic Republic of the Congo). It is a bone tool handle having notches arranged in definite patterns and a bit of quartz fixed in a narrow cavity in its head. It dates back to the period between 9000 B.C. and 6500 B.C. The discoverer of the artifact, Dr. Jean de Heinzelin, suggests that it may have been used for engraving or writing.

He is particularly intrigued by the markings on the bone. There are three separate columns, each consisting of sets of notches arranged in distinct patterns. One column has four groups composed of eleven, thirteen, seventeen and nineteen notches; these are the prime numbers between ten and twenty. In another column the groups consist of eleven, twenty-one, nineteen and nine notches, in that order. The pattern here may be 10 + 1, 20 + 1, 20 − 1, and 10 − 1. The third column has the notches arranged in eight groups, in the following order: 3, 6, 4, 8, 10, 5, 5, 7. The 3 and the 6 are close together, followed by a space, then the 4 and the 8, also close together, then another space, followed by 10 and two 5's. This arrangement seems to be related to the operation of doubling. De Heinzelin concludes that the bone may have been the artifact of a people who used a number system based on ten, and who were also familiar with prime numbers and the operation of duplication.

There is a difference of opinion about the markings on the bone. Alexander Marshack, who has examined the markings on the artifact by microscope, says:

> It represents a *notational* and *counting* system, serving to accumulate groups of marks made by different points and apparently engraved at different times. Analysis of the microscopic data shows no indication of a counting by fives and tens but that the groups of marks vary irregularly in amount. That this is an early system of notational counting is clear; however, this does not necessarily imply a modern arithmetic *numerical* system. I have tracked the origins of such early notational systems back to the Upper Paleolithic cultures of about 30,000 B.C.

In *The Roots of Civilization* Mr. Marshack gives his conclusions (page 364):

> When I examined this tiny petrified bone at the Musée d'Histoire Naturelle in Brussels, I found that the engraving, as nearly as microscopic examination could differentiate the deteriorated markings, was made by thirty-nine different points and was notational. It seemed, more clearly than before, to be lunar.

He plotted the engraved marks on the Ishango bone against a lunar model, and noted a "close tally between the groups of marks and the astronomical lunar periods" (page 30). "A number of other readings in the long series of tests that were conducted gave even closer lunar approximations" (page 31). Here, then, is possible evidence of one of man's earliest intellectual activities, sequential notation on the basis of a lunar calendar, comprising a period of almost six months.

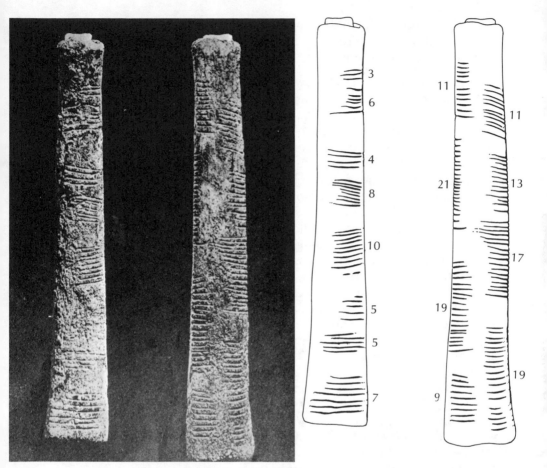

Figure 2–1   Two views of the Ishango bone, over eight thousand years old. This carved bone, discovered at Ishango, on the shore of Lake Edward in Zaire (Congo), indicates that a calendrical or numeration system was known to the fishing and hunting folk of the area. Courtesy of Dr. J. de Heinzelin.

Figure 2–2   Schematic drawing of the Ishango bone, showing the arrangement of notches visible to the naked eye. They suggest a knowledge of multiplication by two and of prime numbers. Microscopic examination revealed further detail.

It is hardly likely that this artifact was unique in the culture of the Ishango area. The newly evolved techniques of microscopic analysis may reveal other such examples. It is also possible that tallies of this sort were made in a more perishable form—on wood, on animal skins, or by an accumulation of pebbles or seeds; if this was so, the evidence would have been lost in the course of time.

When did man first have occasion to keep mathematical records? Was it to note the passage of time, to predict the season for planting the seed, the flooding of the rivers, the coming of the rains? The first calendars, notches on a bone, were probably lunar, following the phases of the moon. But an agricultural economy required the development of a solar calendar and the reconciliation of solar and lunar time-keeping methods. We believe that tallies of the solar year began several thousand years ago in Egypt, and somewhat earlier in Mesopotamia. From the development of the calendar arose the science of astronomy.

The man who herded the cattle and the farmer who cultivated the fields had to observe the passage of the days and the seasons. But it was a separate class of priests who abstracted the practical knowledge into both a scientific study and a religion. Their observations over a period of many centuries enabled them to foretell the behavior of the seasons and the appearance of the heavenly bodies. This knowledge they kept secret. They were agents through whom the people propitiated their gods to ensure the coming of the rains, the appearance of the new moon—the survival and prosperity of the society. Through religious observances the priests exerted their influence over the populace. Knowledge of natural events enabled the priests to predict and claim credit for their occurrence. Frequently they also held the power to divide the land, to demand tribute from the people, to organize public works, and to build vast monuments to the glory of the gods and the kings. Only a stable society, one that had passed the bare subsistence level, could afford to maintain such a superstructure of unproductive rulers and priests.

Mathematics in Ancient Egypt

In ancient Egypt the flooding of the Nile River necessitated annual redivision of the land. Private ownership of land and the ability to produce a surplus of commodities enabled the owners to exchange their products for their private gain or to store them for future use. Thus arose the need for a system of weights and measures. Mathematical operations of addition, subtraction, multiplication, division, and the use of fractions are recorded in Egyptian papyri in connection with the practical problems of the society.

Their methods of doubling and halving, called "duplation and mediation," were still considered separate operations in medieval Europe. This is how the method is used for multiplication. Let us find the

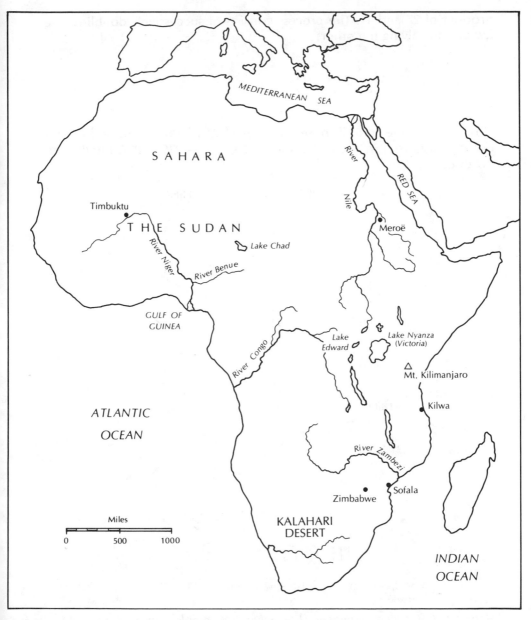

Figure 2–3   Africa: Geographic features and ancient cities.

product of 27 and 11. The process consists of successively doubling one factor and halving the other:

11*	27
5*	54
2	108
1*	216

Now we find the sum of those mutiples of 27 which correspond to odd numbers (starred) in the first column: 27 + 54 + 216 = 297, the desired product.

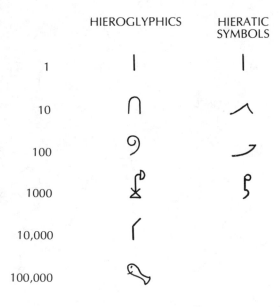

$$3 + (4 \times 10) + (7 \times 100) = 743$$

Figure 2–4 Egyptian numerals. Hieroglyphic symbols appeared as early as 3400 B.C., chiefly in inscriptions on stone. The priests used hieratic symbols when writing on papyrus. The numeral 743 appears in the lower left section of the sandstone stele in Figure 2–5, the account of Amenhotep's expedition. The lower-value signs are inscribed to the left of the higher-value symbols.

Figure 2–5   Egyptian stele giving an account of the expedition of Amenhotep III to Northern Sudan in 1450 B.C. Note the numeral 743, third row from the bottom, left side. British Museum.

The feats of Egyptian engineering are astounding to present-day architects. They were able to construct their pyramids and temples with most minute accuracy. The Egyptian value of π, the ratio of the circumference of a circle to its diameter, was 3.16, much closer to the correct value (3.14159. . .) than was the Babylonian and Biblical approximation of three. They worked out the correct formula for the volume of the frustum of a pyramid (a cut-off pyramid). An early example of rectangular coordinate geometry was discovered in the tombs of the Pharoahs in the form of astronomical texts showing how to consult the position of the stars to determine the time during the darkness of night. This "star clock" indicated the appropriate times for the nightly services in the temples.

Western culture owes a great debt to Egypt and Mesopotamia. The ancient Greeks have been regarded as the fathers of Western civilization. But many centuries before their time the Egyptian priests had developed a complete curriculum for the training of their members. This included philosophy, writing, astronomy, geometry, engineering, and architecture. Indeed, the upper-class Greeks completed their education by studying with Mesopotamian or Egyptian teachers. The Hellenic astronomers adopted the Egyptian civil calendar, the first of its kind in human history. The year consisted of twelve months of thirty days each plus five additional days. (The Egyptians also used the lunar calendar to set their religious observances. This is still done today in computing the dates of the Christian Easter and the Jewish Passover.) The extensive Egyptian libraries were available to visitors. No doubt a great deal of the learning of centuries was written down by the Greeks. Thus they have been credited with discoveries that they merely transmitted from Egypt and the East.

Figure 2–6 Egyptian measuring vessel, marked 8 1/6 hins. About 1250 B.C. British Museum.

Detail of Figure 2–6.

But intellectual stagnation had come to Egypt before the time of the Greeks. Compared with Mesopotamia, the commercial center of the ancient world, Egypt developed in isolation. Only a restricted upper class enjoyed the benefits of the small amount of surplus wealth that this agricultural society could produce. Increasingly, the aristocrats appropriated this wealth for their private enjoyment and for the construction of magnificent monuments in the form of temples and tombs. The result was the virtual enslavement of the poor peasants and craftsmen. The conflicts among various dynasties of kings and between the royal and the priestly segments only weakened the whole structure of the state and led to further decline.

What has happened to the enormous accumulation of literature, science, and technical achievements? Much has been lost; on the other hand, archaeology is revealing to us how much has been assimilated and taken over by later societies. The Bible contains selections which we now realize were taken almost word for word from the Egyptian Mysteries of the Dead. Hellenic science arose on the foundations of several thousand years of accumulated knowledge, and its influence was to be dominant from India to Western Europe until the European Renaissance.

African Glory

Thus far there has been very little effort to investigate the two-way influence between Egypt and the lands to the south and west in Africa. Until recently students of Egyptian history and archaeology have refused to include Egypt and the Nile Valley as an integral part of the African continent. Indeed, the ancient Egyptians were classified as a "Hamitic" branch of the "white race," in spite of ample evidence that blacks played a prominent role in their society. Fortunately the racist "Hamitic" myth has been exploded (see Greenberg).

Our knowledge of the ancient African societies neighboring Egypt is sketchy at best. The kingdom of Kush arose on the upper Nile River, an area that is now in the republic of Sudan. For a time during the first millennium B.C., Egypt was ruled by dynasties from the land of Kush. Its temples and pyramids still stand, but its writing has yet to be deciphered. The principal city, Meroë, was perhaps the first iron-working center of Africa.

With iron, Africans developed new farming techniques. This in turn permitted the accumulation of food surpluses, the maintenance of specialists, such as ironmongers and priests, the establishment of permanent communities, and trade with other peoples. Evidence uncovered by archaeologists has established that iron-working was known in many parts of Africa two thousand years ago.

In West Africa an independent agricultural complex may have evolved at approximately the same time as Egyptian civilization. Cotton and kola, the ingredient of our cola drinks, originated in this region. Murdock, remarking on the inadequacy of our knowledge about the early history of West Africa, says that the archaeologist "has thus far lifted perhaps an ounce of earth on the Niger (River) for every ton carefully sifted on the Nile" (page 73).

The political systems which the Africans developed for governing themselves ranged from centralized states to totally stateless societies. There were many complex political entities in which institutions were developed to make laws, impose taxes, and carry out the functions of a state. Generally speaking, there is a correlation between centralized political power and an advanced economy; there are also remarkable exceptions to this generalization.

A concentration of centralized states arose in the Sudanic area, between the Atlantic Ocean and the Nile, from the desert to the rain forest of the Guinea Coast. We have the greatest knowledge of Ghana, Mali and Songhai, derived from the written accounts of Arab and North African travelers through these lands. The Soninke kingdom of ancient Ghana was already known to the Arabs of the eighth century as "the land of gold." The Soninke were the middlemen in the exchange of gold, panned in the area south of their land, for goods, principally salt, brought across the desert by Arab traders.

By the eleventh century its sources of gold and trade were drying up, and ancient Ghana declined. To the east emerged the empire of Mali, ruled by African converts to Islam. The Arab traveler Ibn Battuta wrote of them in the fourteenth century: "They are seldom unjust, and have a greater dislike of injustice than any other people... There is complete security in their country. Neither traveler nor inhabitant in it has anything to fear from robbers or men of violence." At this time Mali, with its fabulous cities, Timbuktu and Jenne, was one of the largest empires in the world, yet Europeans now found Mali on their maps for the first time.

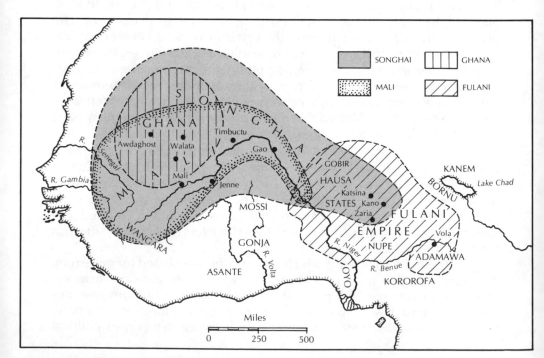

Figure 2–7   Old Sudanic kingdoms.

Mali's absorption by Songhai created an African Muslim empire that was as large as all fifteenth-century Europe. Its political system was based on the feudal relations of land ownership and cultivation. Under the rule of Askia Muhammed Touré, a former slave, ministries were set up to protect the environment, regulate agriculture, finance the empire, and maintain both a standing army of cavalry and foot soldiers and a large fleet of canoes on the Niger River. A unified system of weights and measures was instituted for all markets in the empire and enforced by a network of inspectors. But despite these efforts toward uniformity, the diverse peoples of the empire retained their tribal customs and religious beliefs, little affected by the practice of Islam in the royal court and the towns.

An entirely different structure of society is exemplified by the Igbo of southeast Nigeria, one of the most fertile and densely populated regions of Africa. Archaeological finds indicate that Igbo culture is at least a thousand years old. The Igbo people have always enjoyed a reputation for commercial enterprise, for their profound distrust of authority, and for their eagerness to try anything new. Few kingships developed in their territory. Government was by the village council of elders, but when important decisions were to be made, all the men in the village participated. Since decisions had to be unanimous, the young men were in a position to block actions they found objectionable. Igbo women did not participate in local government, but they have long occupied a prominent position in the commerical life of the society.

By the sixteenth century, if not earlier, the Igbo had a complex economy in which a division of labor was able to support markets every four or eight days. About 1700 the slave trade was assuming a dominant position in the economies of all the peoples along the Guinea Coast and in the Niger Delta. With the help of European guns and bribed with alcohol, the coastal peoples served as middlemen in obtaining inland captives for the European slavers. Here religion and commerce worked hand in hand, just as in other parts of the world. The traders of the Arochuku area had become famous for their services to the Igbo oracle, the Ebinokpapi. Captured men, women, and children were sacrificed to appease the powerful spirit, whose displeasure could cause general disaster in the whole land. Once the captives were offered to the oracle, they were promptly delivered to the waiting European slave ships.

There are no written records to tell the early history of the Central and Southern African states. The Zimbabwe ruins, in present Rhodesia, stand as a testimonial to their forgotten culture. The earliest of these tremendous stone structures was built a thousand years ago, probably by Bantu-speaking peoples. About the fifteenth century Zimbabwe was part of the African state of Monomotapa, whose domains extended from the

Indian Ocean westward to the Kalahari Desert. The gold mines of Mono-motapa, which mining experts estimate had yielded fifteen to seventy-five million pounds sterling, were the principal contributor to the fabulous wealth of medieval India. Unfortunately, from 1896 to 1901 the Ancient Ruins Corporation, a British enterprise, looted the site of Zimbabwe, melting down the ancient gold artifacts and destroying valuable historic evidence.

These few examples of African societies can only suggest the extensive and varied historical backgrounds of the African peoples. The sections of this book devoted to East Africa and southwest Nigeria will treat the political structures and economies of the centralized states, trading cities, and village units of those areas.

SECTION 2
NUMBERS—WORDS, GESTURES, SIGNIFICANCE

The development of numeration in any particular society ultimately rests upon the economic development of the society. African systems of numeration varied from a few number words to well-constructed systems in which counting extended into the millions. Accompanying the number words in many cultures was a formal system of gesture counting. Africans shared with other peoples of the world their belief in the special significance—for good or for evil—of certain numbers.

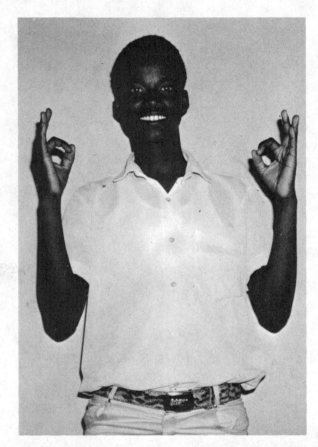

Figure 3–1A   Gesture for "six," Rwanda and western Tanzania. Gesture counting is an integral part of all transactions in the market place; as the number is spoken, the accompanying number of fingers is virtually shaken out with a snap of the wrist. "One" is denoted by extending the forefinger, "two" by two fingers, and "three" by a gesture similar to that for "six", but on one hand only. "Five" is indicated by a closed fist, with the small finger or thumb left out.

Figure 3–1B   Xhosa woman, South Africa, indicating by a gesture that she has seven children. Courtesy of O. F. Raum.

CHAPTER 3
CONSTRUCTION OF NUMERATION SYSTEMS

The impression, among Americans, of African counting is derived from such accounts as the famous sheep-trading story, first told by Sir Francis Galton, the British scientist, explorer, and anthropometrist, in his book *Narrative of an Explorer in Tropical South Africa*, published in London in 1889. The Damara people of Namibia (Southwest Africa) wandered in nomadic bands, grazing their livestock. Here is the tale, as retold in Howard Eves' *In Mathematical Circles* (9°):

> ... the primitive Damaras of Africa, in bartering two sticks of tobacco for one sheep as the rate of exchange, became hopelessly confused when a white trader, desiring two sheep, offered four sticks of tobacco at once. Fraud was suspected by the Damaras, and the transaction had to be revised and carried out more slowly. First two sticks of tobacco were given and one sheep driven away, then two more sticks of tobacco and the second sheep claimed. When shown that the result came out the same as the trader's original proposal, the tribesmen regarded the trader as one possessed of magic powers.

Eves concludes the narrative with a telling comment that is omitted from most other histories of mathematics:

> Yet, these Damaras were not unintelligent. They knew precisely the size of a flock of sheep or a herd of oxen, and would miss an individual at once, because they knew the faces of all the animals. To us, this form of intelligence, which is true and keen observation, would be infinitely more difficult to cultivate than that involved in counting.

There is ample evidence in literature of the Africans' skills in the use of numbers. A British trader named Clapperton, reporting in 1826 on the use of cowrie shell currency in the Niger River region, praised the "dexterity of the natives in counting the largest sums" (Hodgkin, page 215). In *Primitive Money* Paul Einzig states that merchants had to count in large figures, and they developed the advanced arithmetical faculty required for the purpose. Michael Crowder, in *West Africa under Colonial Rule*, speaks of the Yoruba people. Part of the culture that each generation transmitted to the succeeding generation was the ability to handle the various kinds of currency then in use. Since a low-valued cowrie currency was used as the base, considerable arithmetic skill was necessary to conduct trade outside of the confines of the village. "Contrary to a generally European held opinion that Africans could not count beyond ten, the Yorubas could count to a million" (page 373).

The descendants of slaves brought to the coastal areas of South Carolina and Georgia during the eighteenth and early nineteenth centuries still speak Gullah, a combination of English and the West African language of their ancestors. Among the Africanisms surviving to the 1940s, according to Lorenzo Turner's study, were the numerals from one to ten in several West African languages, and from one to nineteen in the language of the Fulani people. The survival of these number words shows that the numerals constituted an important part of the vocabulary of the slaves who brought their languages from Africa.

An excellent booklet by Gay and Cole describes the Kpelle people's experience with numbers and measurement. Most arithmetic operations are performed with the aid of piles of stones. Consequently, the people become extremely adept in recognizing numbers in this way. In fact, in an experiment requiring estimation of the number of stones in piles of various sizes, nonliterate Kpelle adults achieved far better scores than did Yale undergraduate students. They also excelled in estimating the amounts of rice in containers; the activity of measuring rice was an integral part of their lives. But when asked the number of people or houses in their village, the Kpelle subjects did poorly. Perhaps they were inhibited by fear of the tax-collector!

How Do We Count?

When we speak of our numeration system, we may be referring to either of two distinct ideas: a set of written symbols or a collection of number words. When an African speaks of his traditional numeration system, he may mean a collection of number words or a set of standardized gestures which accompany or replace the words.

The Meaning of Base

The written symbols by which we represent numbers enable us to express in simple notation the distance to a star or the size of an electron. These Hindu-Arabic numerals require just ten distinct symbols: 0, 1, 2, 3, 4, 5, 6, 7, 8 and 9, and that marvelous invention called "place value." By means of the arithmetic operations of addition and multiplication on the ten fundamental symbols, a number of any magnitude can be expressed. The operations are implicit rather than clearly expressed. It is the principle of place value that enables us to write the symbols 23 to denote $(2 \times 10) + 3$. We say that our system is based on ten, and refer to it as a decimal system, since each position in a written numeral has a value ten times that of the position to its right.

For example, analyze the meaning of the numeral 4306. Implicit are the operations:

$$(4 \times 1000) + (3 \times 100) + (0 \times 10) + (6 \times 1).$$

The place value principle is incorporated in the abacus (Figure 3–2).

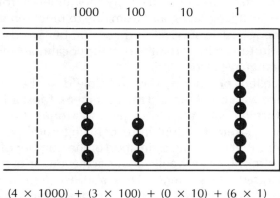

1000   100   10   1

$$(4 \times 1000) + (3 \times 100) + (0 \times 10) + (6 \times 1)$$

Figure 3–2   Abacus based on ten.

Many thousands of years elapsed from the time man began to count to the development of this very convenient method of expressing numbers. Within our own English number words are contained the vestiges of primitive ideas of counting. The word *three* has been thought by some historians to be related to a word for "beyond" (more than one can count), and to have meant "many" (Menninger, page 17). There is an obvious relationship between the words *three* and *third*, and between *four* and *fourth*, but where is the connection between *one* and *first*, or *two* and *second*? The words *one* and *two* are retained from a period before the invention of the counting sequence.

Analyze the number words after *ten*. *Eleven* means "one left," an abbreviation of "ten and one left over," while *twelve* is literally "two left." No doubt the terminology refers to the additional fingers needed after having used all ten fingers, or digits, on both hands. To continue with the number sequence, *thirteen* means "three and ten," a formulation based on the addition of digits to *ten*. When we come to *twenty*, we are using the multiplication process to express "two times ten"; *thirty* means "three times ten," and so on to *ninety*. *Hundred* is derived, by a process of many linguistic changes, from words denoting "ten of tens" (see Menninger). The multiplication of each rank to form the next may be explicit, as in *ten thousand*, or implied, as in the word *million*, from the Latin *mille*, meaning "thousand."

In many languages the number words are constructed with five as a primary base, no doubt in relation to the five fingers of one hand. In such cases there is a secondary base, usually ten or twenty.

The Japanese abacus, called a soroban, has as its basic numbers both five and ten. In each column there are four counters in the lower section and one in the upper section. A number from one to four is indicated by pushing up the appropriate number of lower beads. Five is indicated by pushing down the solitary upper bead. A number from six to nine is regarded as a sum of five and an appropriate smaller number. Each column represents a value ten times that of the column to its right. Thus, the number represented by the symbols 10,653 is indicated on the soroban shown here.

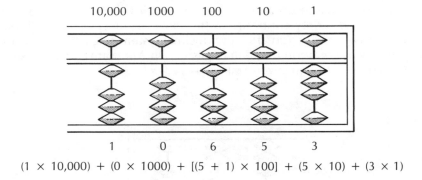

$$(1 \times 10,000) + (0 \times 1000) + [(5 + 1) \times 100] + (5 \times 10) + (3 \times 1)$$

Figure 3–3   Soroban, based on five and ten.

In the linguistic sense we shall use the word *base* to indicate the number or numbers which dominate the construction of the number words. The base is used in composition with other number words, either implicitly or explicitly, to build the number sequence.

The majority of African numeration systems are quinary; that is, they have five as their primary base. The words for two, three, and four have generally survived from the earliest existence of the tribe, since a bare subsistence economy required little more in the way of counting. Indeed, these are among the words used to classify the hundreds of African languages. As the economy grew, so did the numeration systems. Some peoples adopted ten as a secondary base while others used twenty. Nowhere do we find gradations based solely on five and powers of five (for example, five times five). This is perfectly natural, for the process of enumeration evolved as a one-to-one correspondence with parts of the body, in some cases the fingers, in others both the fingers and toes. But let us not be too literal. We cannot conclude that counting was actually carried out on the fingers or on the fingers and toes, depending upon whether ten or twenty was used as the base. Few African peoples count on their toes. The Eskimos, who are among the many peoples of the world who developed a twenty-system, do not have easy access to their toes! The concept implied by the number word for twenty in some languages—"man complete"—is that each object to be counted is paired off implicitly with a digit (finger or toe) of a person's body.

Counting based on five and twenty, called quinquavigesimal counting, is widespread throughout the world. The Aztecs and Mayas of ancient Mexico used this system. The French number sequence proceeds smoothly by tens until it reaches seventy, *soixante-dix*, or "sixty-ten." Eighty is clearly based on a gradation by twenties—*quatrevingt*, or "four twenties"—and ninety continues in the same manner, by the addition of ten to four twenties. Thereafter the decimal system rules. It is not so long since *score* was common usage in English, as in Abraham Lincoln's famous words, "Four score and seven years ago...."

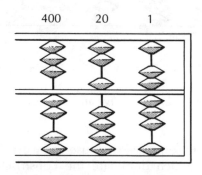

$$(2 \times 400) + (5 \times 20) + [(5 + 5 + 3) \times 1] = 913$$

Figure 3–4 Soroban, based on five and twenty. Representation of the number 913 on a non-existent soroban-like abacus. The four lower beads in each column represent units, while each bead in the upper section represents five. Each column has a value twenty times that of the column to its right.

Special usage, most commonly currency denominations, introduces bases other than those employed in the system of number words. We have in the English language many vestiges of principal numbers different from ten; twelve units in a foot or a dozen, sixty in time and angle measurement, twenty in a score. In Africa, cowrie shell currency was associated with many sets of special numbers.

The operations of addition and multiplication, and in some cases of subtraction and doubling, on the basic numbers produce larger numbers. Generally, the African number words express the operations explicitly—"five and one," "twenty three times," "twenty take away one," "four-four."

Finger Gestures

We have noted vestiges of finger counting in our words *eleven* and *twelve*. The number words in some African languages clearly express the finger gestures. In Zulu, the word for six means, literally, "take the thumb," indicating that all the fingers on one hand have been counted and one must start with the thumb of the second hand.

On the association of parts of the body with number words, Frank Chapman writes in *Science and Africa*:

> I don't think we can say with certainty at just what stage in the historic process man began to think of abstract numbers. When I say something is twenty feet I don't think of twenty human feet; by the same token when the Zulu says, "taking the thumb," I don't think he's necessarily thinking of his thumb . . . I wonder how many anthropologists take the expression of these people literally?

There are many African societies in which the finger gestures have equal status with the spoken numerals and consitute a proper system of numeration which may or may not agree with the spoken number words in its derivation.

Recorded instances of finger counting go back to ancient Egypt. In a pyramid text, the *Book of the Dead*, the deceased king tries to engage a ferry-boat to take him across a river in the lower world. Neugebauer relates the story (page 9):

> The ferryman objects: "This august god (on the other side) will say, 'Did you bring me a man who cannot number his fingers?'" But the deceased king is a great "magician," and is able to recite a rhyme which numbers his ten fingers. . . . This act had magical significance.

Expansion and Modernization of the Number System

As its economy grew more sophisticated, a people naturally required numerals of a higher order. In some cases there was borrowing from neighboring languages, in others the additional words are indigenous. Since most African societies had no written records, it is often difficult to trace the history of the numerals. We find the Arabic word *alif*, the first letter of the alphabet, used for "one thousand" in some Sudanic languages, and in modern Mende "one hundred" is *hondo*, from the English. Words of alien peoples enter a language because of commerce, the widespread shifting of populations, or the conquest of one people by another. To quote W. L. Migeod: "This war of the numeral system is the sole historical record left by antiquity of raids, wars, and of much misery to the human race. The numerals are the historians of those bygone conflicts, and who can say the price in blood of one strange word or particle found in a language?" (page 116).

In the course of the years the number words or their meanings undergo alteration, due to reasons other than the expansion of the numeral system. Special number words associated with the counting of particular objects, such as cowrie currency, tend either to disappear from use, as the needs of the society change, or to assume a more general application. Furthermore, when a written language is developed, there is standardization of the various words for a given concept in the different dialects of a given language. Reducing the length of an expression for a number is an important objective in a society that strives to spread formal education among the people.

To illustrate streamlining, Cecilia Irina told me that the cumbersome Sotho construction for ninety-nine:

> *mashome a robileng meno o le mong a (nang le) metso e robileng mono o le mong*, literally "tens which bend one finger which have units which bend one finger"

has been replaced by:

> *mashome a robong a metso e robong*, "tens nine with a root that is nine."

CHAPTER 4
HOW AFRICANS COUNT

Numbers One to Nine

In spite of the wide distribution of African peoples and the existence of perhaps a thousand languages on the continent, the words for two, three and four are similar in an area covering about half of Africa. This area includes the Sudan—the region extending from the Sahara southward to the Gulf of Guinea, and from the Atlantic Ocean to the Nile River—and most of the southern part of the continent, now inhabited by Bantu-speaking peoples. The three hundred Bantu languages are classified by Greenberg as a subgroup of the Niger-Congo family, and bear a great resemblance to the languages of the western Sudanic peoples with respect to certain basic words. Linguists conclude that both categories of languages had a common origin, and that the Bantu-speaking peoples dispersed throughout the southern half of Africa from the Nigeria-Cameroons area.

The word for one exhibits great variety in African languages, as the reader will note from an examination of the lists of number words appearing in the text. Not so the words for two, three and four. "Two" is usually a form of *li* or *di*. The word for three contains the syllable *ta* or *sa*, and "four" is generally a nasal consonant, like *ne*. "Five" has a variety of forms; frequently it is the word for hand.

Dr. Schmidl introduces her discussion of Bantu numeration with these words (page 168):

When we compare the number words from one to nine in the various Bantu languages, we find a similarity in the names for 2, 3, 4, and 5, while the corresponding gestures differ considerably. The basic stems are:

        2   *-vili* or *-vali*
        3   *-tatu*
        4   *-na*
        5   *-tano*

... There are various expressions for "one", but generally they are related to *-mwe*.

In contrast, there are wide variations in the words for 6, 7, 8 and 9, and it will be necessary to deal separately with the various branches of the Bantu languages.

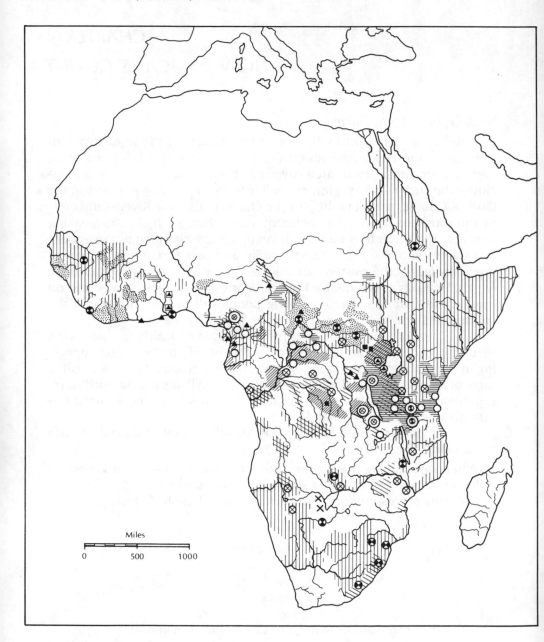

Figure 4–1   Distribution of numeration systems in Africa (after Schmidl).

Key for Map, Figure 4–1.

Distribution of Numeration Systems in Africa (after Schmidl)

Code

I. Verbal numeration systems

   A. Formation of numerals for 6, 7, 8 and 9 in:

     1. Base-ten systems

       a. On the principle of two approximately equal terms

       At least two numbers

       Just one number

       b. By composition with five

       At least two numbers

       Just one number

     2. Base-twenty systems

       Formation of at least two numbers by composition with five

   B. Unusual bases

     1. Evidence of base-two counting

     2. Composition with six

     3. Composition with eight

II. Gesture counting

   A. In the manner of the Shambaa

   B. In the manner of the Hima

   C. In the manner of the Zulu

   D. Based on five and ten

   E. Based on five twenty

It is precisely the words for 6, 7, 8 and 9 which exhibit the most intriguing constructions. In some cases we find a simple addition to five; for example, in Kwanyama (southwestern Africa):

    6 tano-na-mwe
    7 tano-na-vali
    8 tano-na-tatu
    9 tano-na-ne

The composition of one with five to express "six" may not be so obvious. In the Malinke language, spoken on the upper Niger River, writes Maurice Delafosse, "six" is expressed by *woro* or *wolo*, where *wo* is an abbreviation of the word for five, and *ro* or *lo* is a shortened form of "one."

In many languages the words for six through nine are derived directly from the gestures for these numbers, and are based on several different systems of gesture counting. This topic will be discussed fully later in this chapter in connection with gestures.

Remarkably, the Malinke word for nine, *kononto*, means literally "to the one of the belly," a reference to the nine months of pregnancy! This is obviously a word of the common people.

Five-Ten Systems

The Sudanic and Bantu branches of the Niger-Congo language family diverge on the choice of the secondary base, the former generally using twenty, and the latter favoring ten.

Most Bantu languages use *kumi* or *longo* for ten, although that is not precisely the original meaning. The higher decades proceed smoothly as multiples of ten.

Number words for a hundred, a thousand, and higher ranks were rarely used except in association with specific objects to be counted. In the Ziba language:

100	*tsikumi*, referring to a string of 100 cowrie shells
1000	*lukumi*, a bundle of ten strings of cowrie currency
10,000	*kakumi*, a heap of ten bundles

As the need arose, these same words were applied to other objects, or new words were invented.

In the section devoted to East Africa, we shall discuss some of the Bantu languages in greater detail, as well as the Nilotic languages – the Maasai, the Luo, the Kalanjin. In the latter the numeration systems are based on ten, or on five and ten.

Five-Twenty Systems

In the languages which build on twenty as a secondary base, "ten" may be an independent word, or it may mean "two hands" or "two fives." It is followed by an expression denoting "ten and one," and so on to "fifteen," which may be "two hands and one hand," or "ten and one hand," or even an independent word, as in the Dyola language of Guinea-Bissau, where the word means "to bow." Continue by adding one, two, etc. to the word for fifteen, until we reach twenty. This word in some languages means literally "man complete." In the Banda language of Central Africa the word for fifteen means "three fists," and for twenty, "take one person." The same method is then used all over again, adding to the word for twenty to form the numerals from twenty-one to thirty-nine. Forty is expressed as "two men complete," and one hundred is "five men complete"; in other words, all of the digits on the hands and feet have been counted five times. One Malinke expression for "forty" is *dibi*, "a mattress," the union of the forty digits, "since the husband and the wife lie on the same mattress and have a total of forty digits between them," to quote Delafosse (page 389).

This is the basic form of a quinquavigesimal system (based on five and twenty), in which five is the primary base. It is found in many languages of the Sudan region: Dyola, Balante and Nalu in the west, Yoruba, Nupe and Efik in Nigeria, Vai and Kru in Liberia, as well as the Nuba language of the eastern Sudan.

As a numeration system develops, special words are introduced, or words take on a new meaning. Foreign influence may bring about linguistic changes. The Malinke word *keme*, meaning "a large number," came to denote "one hundred," and was used in counting cowrie currency instead of the more cumbersome expression "five men complete."

The number words of a language may not be the same in sources from different periods. Migeod stated that the Mende say "five men finished" for one hundred, but a recent instruction book on the Mende language gives *hondo yila*—the word *hondo* is derived from the English *hundred*, and *yila* means "one."

The Igbo numeration system is based on twenty; whether five or ten is also a base is not so apparent. Thirty is expressed as the sum of twenty and ten, fifty is forty and ten, and similarly for the larger numbers (Table 1). In most dialects the formation of the numbers in any decade is by addition of the digits from one to nine to the appropriate word for the multiple of ten. Abraham does give an alternative method: 16, 17, 18 and 19 can be formed by deducting 4, 3, 2 and 1, respectively, from 20, and similarly for the higher decades. None of my informants could confirm this construction. Nor could I discover any basis for the statement by Thomas, a British anthropologist of the early twentieth century, that the Igbo people use 25 as a base, except for the counting of cowries.

There is a special word for the square of twenty, as one would expect in a vigesimal system—four hundred is *nnu*. The square of 400, or 160,000, is expressed as *nnu khuru nnu*, literally "400 meets 400." An "uncountably" large number is *pughu*.

A unique system, based on twenty, is that of the Yoruba people of southwest Nigeria. Yoruba numeration illustrates an unusual subtractive principle still in effect today. The number forty-six, for example, is literally "twenty in three ways less ten less four," or $(20 \times 3) - 10 - 4$. This system will be treated more fully in the chapter on mathematics of southwest Nigeria.

Along the Niger River, for several hundred miles from the coast, the subtractive method of forming number words is common. Perhaps this

1	otu	30	ohu na iri
2	abuo		(20 and 10)
3	ato	31	ohu na iri na otu
4	ano		(20 + 10 + 1)
5	iso	40	ohu abuo
6	isii		(20 × 2)
7	asaa	50	ohu abuo na iri
8	asato		$[(20 \times 2) + 10]$
9	toolu	60	ohu ato
10	iri		(20 × 3)
11	iri na otu (10 and 1)	100	ohu iso
12	iri na abuo		(20 × 5)
	(10 and 2)	200	ohu iri
20	ohu		(20 × 10)
21	ohu na otu	300	ohu iri noohu ise
	(20 and 1)		$[(20 \times 10) + (20 \times 5)]$
		400	nnu

Table 1. Igbo Number Words

was the easiest method of counting cowries, or it may be that traders from one ethnic group influenced and to some extent standardized the numeration systems. Within a given language, dialects differ as to whether they compose by addition or by subtraction. Investigators give contradictory word lists, depending upon the dates and areas of their research. With the present trend toward standardization of languages, the discrepancies should be ironed out in time. Literacy and formal schooling require a uniform construction and spelling of the number words. These factors, reinforced by the mass media, tend to obliterate the various shades of local color.

Linguistic Change

The Arabs brought the new religion of Muhammed into North Africa during the seventh century. By the eleventh century there had been a fairly large immigration, and Islam spread slowly southward across the Sahara. Through immigration and commercial contacts, the influence of Islam was pervasive in many large centers of the Sudanic empires. Any person professing to practise the religion was obliged to read the Koran in the original

Arabic. Thus Arabic became the language of the cultural centers, and Arabic numeration replaced or supplemented the indigenous systems. However, outside the commercial and intellectual concentrations, people lived their lives in the traditional way. This may account for the fact that in the Sudan the Soninke, the Hausa, the Fulani, and the Songhai count on the basis of tens, and even use some Arabic words, while their neighbors retain twenty as the secondary base. Further south, near the Guinea coast, we find a decimal system in use among the Ga, the Twi, and the Kpelle.

Arabic influence is apparent in Hausa, the most widely used language in northern Nigeria. Nineteenth century linguists classified the Hausa numerical system as quinary-vigesimal, but having twelve as the base for the formation of the words for thirteen through eighteen. Later sources showed a clear relationship to the decimal Arabic system with respect to the numerals starting with twenty as well as the use of the Arabic word for six. An interesting aspect of the numeration system of some eastern Hausa peoples is the use of a subtraction principle for compound numbers ending in eight and nine; e.g., $18 = 20 - 2$, and $19 = 20 - 1$. However, a recently published Hausa grammar gives a system of numeration that is pure decimal.

Migeod reported three designations for twenty in use at the same time: (1) *hauiya*, a special word applied to twenty cowrie shells, used as currency, (2) the plural of *goma*, the word for ten, and (3) the Arabic word for twenty. Some Hausa dialects had special words for a bag of 20,000 cowrie shells.

These are the Hausa number words in current use:

1	daya	8	takwas
2	biyu	9	tara
3	uku	10	goma
4	hudu	19	goma sha tara
5	biyar	20	ashirin (from Arabic)
6	shida (from Arabic)	30	talatin (from Arabic)
7	bakwai	100	dari

Unusual Number Bases

Most methods of number-name construction in Africa are based on five and have a secondary base of either ten or twenty. The mathematical operations of addition, subtraction and multiplication on the basic numbers give rise to names for larger numbers. Some numeration systems, however, show traces of bases other than five, ten and twenty.

As noted above, the Hausa people of northern Nigeria formerly constructed the numerals from thirteen to eighteen by addition to twelve.

Some distance south of them live several technologically primitive tribes, who also use a base of twelve superimposed on a five-ten structure in which six through ten are formed by composition with five. Although these tribes have different vocabularies up to ten, all their languages have similar words for eleven and twelve, and they all form thirteen as twelve and one, fourteen as twelve and two, etc. One would suspect a common source of the number words. A plausible theory is that these people borrowed from the Hausa, who have for centuries worked the tin mines in their area.

In the languages of the Bram and the Mankanye people of Guinea-Bissau, five is denoted by the word for "hand," ten is "two hands," nine is "hand and hand less one," and nineteen is expressed as "two hands and hand and hand less one," certainly a quinary system. But twelve is "six times two"! Twelve is formed in the same manner by the Bolan of Guinea. They carry the idea further with $24 = 6 \times 4$. Their neighbors the Balante compose with six to form the numerals from seven through twelve: $7 = 6 + 1$, $8 = 6 + 2$, etc. In the Ga language of Ghana, both seven and eight are based upon six: $7 = 6 + 1$ and $8 = 6 + 2$.

The Huku language of central Africa displays the greatest variety, basing many names on four and six:

$7 = 6 + 1$	$13 = 12 + 1$
$8 = 2 \times 4$	$16 = (2 \times 4) \times 2$
$9 = (2 \times 4) + 1$	$17 = (2 \times 4) \times 2 + 1$
10 is an independent word	$20 = 2 \times 10$

Gesture Counting

You may have ticked off on your fingers the number of guests you expected or the number of days to the next holiday. Perhaps you started with your thumb or maybe it was your little finger. Did you think about whether you used your right hand or your left? Did it matter whether you extended each finger, bent it, or tapped it with the index of the other hand?

The African would no more think of using his fingers so haphazardly than you would count: "three, one, six, eleven. . . ." To the African the fingers gestures constitute as formal a method of counting as do the spoken words. Most often the two methods of expression are used simultaneously—the gestures accompany the spoken words. Go to a Hausa market and you will see people bargaining about prices, to the accompaniment of finger gestures. To indicate five, one hand is raised with the finger tips bunched together, and for ten, the hands are brought together, with all the fingertips of both hands touching.

Our two hands, a built-in calculating machine, gave rise to a decimal system of counting in many European languages. The intimate relationship between number words and finger counting is even more apparent in many African languages.

Consider the Zulu numerals, and their relation to finger counting. The finger gestures start with the left hand; the palm is up and the fingers are bent:

Numeral		Derivation	Finger gesture
1	nye	State of being alone	Extend left small finger
2	bili	Raise a separate finger	Extend left small and ring fingers
3	thathu	To take	Extend three outer fingers
4	ne	To join	Extend four fingers
5	hlanu	(All the fingers) united	Extend five fingers
6	isithupa	Take the [right] thumb	Extend right thumb
7	isikhombisa	Point with the fore-finger of [right] hand	Extend right thumb and index finger
8	isishiyagalombili	Leave out two fingers	Extend three fingers of the right hand
9	isishiyagalunye	Leave out one finger	Extend four fingers of the right hand
10	ishumi	Cause to stand	Extend all fingers

There is no doubt about the origin of the number words in Zulu. It is true that some words are long and cumbersome, but the efforts of the school authorities to shorten them have met with little enthusiasm.

The Sotho of southern Africa say tselela or tsela for "six," a word derived from the expression "to cross over." The reference is to the gesture language, crossing over from one hand to the other, a construction similar to that of Zulu. To continue with the Sotho numerals:

7   supa, "point"
8   robeli, "bend two"
9   robong, "bend one"

again similar to the Zulu expressions.

Some Bantu and Sudanic languages use words meaning "three and three" for six, "four and three" for seven, "four and four" for eight, and "five and four" for nine. Especially common in the Bantu languages is a word similar to *nana* for eight that is derived from the stem *-na*, meaning four. The origin of this compounding of words lies in the method of finger reckoning called "representation by two approximately equal terms."

The finger gestures are:

6   three fingers of each hand
7   four fingers of one hand, three of the other
8   four fingers of each hand
9   five fingers of one hand, four of the other

A good example is the Ekoi language of Cameroon:

3	*esa*	6	*esaresa* (*esa* and *esa*, or 3 + 3)
4	*eni*	7	*eniresa* (*eni* and *esa*, or 4 + 3)
5	*elon*	8	*enireni* (*eni* and *eni*, or 4 + 4)
		9	*eloneni* (*elon* and *eni*, or 5 + 4)

This variation is widespread east of Lake Tanganyika, through the upper Congo River area, in Cameroon, near Lake Chad, and in South Kordofan. The Felup of Senegal use a combination of two methods:

$$6 = 5 + 1 \quad \text{(quinary)}$$
$$7 = 4 + 3 \quad \text{(two approximately equal terms)}$$
$$8 = 4 + 4$$
$$9 = 5 + 4$$

Names for the numbers from six to nine in some languages are based on the simple combination of the fingers of one hand to all five of the other. The Herero (Bantu language of southwest Africa) word *hamba* means "to change over" to the other hand when counting on the fingers. The number words are:

6   *hamboumwe*, from *hamba* and *umwe* (one)
7   *hambombari*, from *hamba* and *imbari* (two)
8   *hambondatu*, from *hamba* and *indatu* (three)
9   (*hambo*) *muviu*, from *hamba* and *imuviu* (unexplained)

The corresponding gestures are:

6   place little finger of right hand on left thumb
7   place two outer fingers of right hand on left thumb

8 place three outer fingers of right hand on left thumb
9 place four fingers of right hand on left thumb

We can now present some generalizations about finger counting. In those systems that build by addition to five, counting usually starts with the little finger of one hand and proceeds by the addition of the appropriate fingers in sequence until five is reached. This number is generally denoted by a closed fist. For six, the little finger of the other hand joins in the counting, and the fingers of the second hand are used in the same sequence as those of the first. Exceptions to this generalization are the Zulu and Sotho systems, where counting from six to nine commences with the thumb of the second hand and proceeds towards the little finger. The fingers may be extended, as with the Herero and Zulu, or bent down, as with the Malinke and Ewe of the western Sudan. This type of gesture counting is called quinary, since it is based on five.

In systems that are dominated by the principle of two approximately equal terms, counting usually starts by extending the index finger, and proceeds by using an additional finger in sequence for each number until the four fingers of one hand have been extended. The gesture for four is also based on this method of duplication, in that the first and second fingers are separated from the third and fourth. Five is again a closed fist.

Gesture counting will be considered in greater detail in the chapter on East African numeration, where several Bantu and Nilotic systems will be discussed and compared.

We have seen that number words may be derived from gesture language, as in the Zulu, Ekoi and Herero languages. By no means do the two always coincide; the contrary is often true. We may find pure quinary gestures accompanying verbal expressions based on two approximately equal terms, or vice versa. In fact, historians and anthropologists use both the oral and the gesture languages of a people to trace their dispersal patterns and their relationships with other ethnic groups.

A case in point is the method of representation by two equal terms, based on the operation of doubling, or duplication. It brings to mind both the Egyptian method of multiplication and division by duplation and mediation (see Chapter 2) and the ancient Egyptian hieroglyphic numerals. On many monuments and temple walls we find:

‖	⫼	⫼	⫼⫼	⫼⫼		⫼⫼
‖	‖	⫼	⫼	⫼⫼	but	⫼⫼
4	5	6	7	8		9

The later hieratic symbols are:

Ш or —   ᵞ or 7   ⁄⁄ᵞ   ⸝ or 2   ⇉   ʒ

4   5   6   7   8   9

Figure 4–2

These number symbols, and the fact that the Egyptians used finger gestures based on this principle (they also had quinary finger counting), confirm the prominence of multiplication by two in the development of Egyptian numeration. Going back several more thousands of years, one finds that the Ishango bone, with its sequence of notches, also suggests a notational system based on doubling: 3 followed by 6, 4 followed by 8, then 10 followed by 5 and 5 (see Chapter 2). It is conceivable that the system spread from Ishango northward down the valley of the Nile. Is it just a coincidence that this bone was found on the shores of Lake Edward, in the heart of the region where counting according to the principle of two approximately equal terms prevails today? Perhaps further archaeological research will give the answer.

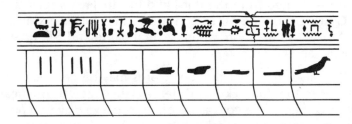

Figure 4–3   Egyptian carving showing the measurement of the ell, which contained twenty-eight "fingers." The first three are represented by the corresponding numbers of strokes, followed by finger gestures for four, five, six and seven. Four is a pictograph of a hand with the thumb closed over the palm, five is an open hand with thumb extended, six is a closed fist with thumb extended, but seven is not clearly portrayed.

Gesture language for the higher denominations is illustrated by the Efik, in the lower Niger Valley, who use the following elaborate system, as quoted in Schmidl's article (page 191) from a work by Leonard:

A hand clenched means 5, and along with one up to four fingers, from 6 to 9, both hands being clasped for 10. A finger is again added for 11 and so on up to 15, for which the arm is bent and the hand touches the shoulder. Twenty is signified by waving a finger in front of the body, and the reckoning proceeds as before until 30 is reached, when the hands are clapped and a finger waved. For 40, two fingers are waved, and at 50 the hands are once more clapped, and in the same way the remaining fingers up to a 100 are signalled, when the closed fist is waved and the simple sum of addition comes to an end.

The Ekoi people of Cameroon are among the few African peoples who actually do count on their toes. Both their spoken and gesture languages are based on representation by two approximately equal terms, but they show "one" on the little finger, as do the base-five counters, rather than on the index finger, and their secondary base is twenty, as is common among the Sudanese. This combination of Bantu and Sudanese methods is hardly surprising, since their home is on the border of the two areas.

CHAPTER 5
TABOOS AND MYSTICISM

"Under taboo, if men made no missteps they lived safely, but they had to be continuously alert to the consequences of infringement of the order of nature and the order of social life." So writes Margaret Mead of a pre-industrial society.

Taboo on Counting

There is a widespread fear among Africans that the counting of human beings, domestic animals and valuable possessions will lead to their destruction. To circumvent the taboo, counting is done indirectly, by setting up a one-to-one correspondence with some type of counting device.

Taboos on counting are found throughout the world and throughout the ages. The Jews require that ten men, a *minyan*, be present in the temple for religious services. The word *minyan* itself means "count." To ascertain that the quota has been filled, a ten-word sentence is recited, establishing a one-to-one correspondence between the worshippers in the synagogue and the words of the sentence.

One can recall, too, the Biblical story of King David, who acted against the advice of Joab and the other army captains in ordering a census of his people. After the count had been brought to him, he realized his error, but it was too late. Jehovah visited a terrible pestilence on the people, and many died. Only the offering of a sacrifice succeeded at last in staying his hand.

The superstition about counting is treated at length by A. Seidenberg in his article, "The Ritual Origin of Counting."

Yet a centralized state has to collect taxes, and it therefore needs an accurate count of the number of people living in its domains. This was accomplished in precolonial African states by various forms of indirect census. Herskovits describes the procedures in the highly organized kingdom of Dahomey, where, by the way, the women of the royal court had control over all transactions. On certain market days, a crier would be sent by the chief priest of the powerful spirit of the sacred river, to announce that the spirit threatened disaster to crops and livestock unless the people did as he bade. Every man and woman was to bring to the palace a cowrie shell for each animal he owned, and to deposit the shells in separate piles for sheep, goats, and cattle. First he must touch the animal with the cowrie, to transfer the danger from the animal to the shell. The king contributed an equal number of cowries, and retained a pebble for each shell he contributed. Thus the royal bureaucracy secured an accurate count of livestock, by kind and by village, to be used as a basis for fiscal computations.

Hunters in Dahomey were counted indirectly during the rituals for the gods of the hunt. The heads of all slain animals had to be sent to the palace, thus controlling the amount of game killed. Peasants, ironmongers, weavers, traders—all were assessed a certain part of their products or their sales, thus furnishing a count of the donors and an estimate of the gross national product.

The Feast of the New Yam is the most significant Igbo festival, marking the harvest and the beginning of the New Year. This is the occasion for the annual census. Every grown man brings to the sacred shrine one yam for each member of his household. The number of yams contributed by the village is counted and announced to the assembled populace by the priest. There is great joy if the population has increased.

The colonial invaders had little understanding of and no sympathy with this fear on the part of the African people. No doubt many open struggles were precipitated by the administrators' insistence that they be permitted to count people, houses, or livestock.

Number Symbolism and Superstitions

No phenomenon associated with African numbers has been so misinterpreted by European investigators as the subject of number symbolism and taboos. Frequently the interpretations they gave were their own, not those of Africans. As a result, it is difficult to distinguish between numbers that are associated with the world view of the people, and numbers that are taboo because of superstition. Here I am describing the reports of both European and African informants and relating them to the beliefs that have influenced the development of Western culture.

The Pythagorean school of ancient Greece regarded even numbers as feminine, pertaining to the earthly, and odd numbers as masculine, relating to the celestial. Many Sudanic peoples have similar beliefs about the association of numbers with male and female.

Among the Kolokuma Ijo people of the Niger Delta, odd numbers, especially three, are associated with men, and even numbers, four in particular, with women. Within the four-day week, the first and third are the strong days, favorable to men; the second and fourth are lucky for women. Along similar lines, to refer to a Kpelle man as "four" is to curse him. Among the Dogon, of the middle Niger River region, a woman's cloth is made of four woven strips in token of her femininity. The male number, three, appears in the three sets of three strips each, sewn together to form a man's trousers.

According to Frobenius, the Akan peoples associate the queen mother with "three" and the king with "four." They consider the odd numbers three, five, seven and nine to be favorable indivination. But Rattray reported that the Asante, one of the Akan peoples, consider five a most unlucky number. One would give five of anything only to his worst enemy!

When a person is in trouble, he is likely to seek advice from the diviner—should he leave home to find a job in the city, is his beloved betraying him, will his mother's illness prove fatal? The Igbo occultist's equipment includes four cowries. He rolls out the shells and examines their positions. If all four land with the openings down, the omens are most favorable. Most disastrous is an outcome of three cowries in one position and the fourth in the opposite position.

Figure 5-1 The Igbo diviner bases his predictions on the outcomes when four cowrie shells are cast. Most favorable is that in which all four shells land with the closed part uppermost. An outcome of three shells in one position and the fourth in an opposite position is considered a bad omen.

An integral part of an Igbo wedding is the consultation with the diviner to learn what the future has in store for the young couple. Here again is the symbolic number four, typical of southern Nigeria. Achebe describes the medicine man's ceremonial objects: four small yams, four pieces of white chalk, four palm leaflets. Then the young man's mother counted out four groups of six cowries, as she would at the market, and gave them to the diviner (from *Arrow of God*).

Few European investigators have been able to gain insight into the beliefs of African peoples as did Marcel Griaule, as related in his book *Conversations with Ogotemmêli*. In a series of remarkable interviews the sightless old man of the Dogon people revealed the beliefs of his ancestors. Central to Dogon cosmology is the number eight. In the beginning were the eight ancestors, the progenitors of the eight Dogon clans. The Nummo spirit, the architect of the world order, laid out the eight covenant-stones in an arrangement which at the same time indicated the outline of a human soul and the order of human society. "The *dougué* (covenant-stones) were eight in number like the eight ancestors or the eight sorts of seed. They represented the eight elders at the origin of mankind; and the eight men, the eight seeds and the eight joints are all the same order as the *dougué*" (page 51). In the outline of a man's soul, made at every birth by the Nummo spirit, eight cowries are put in place of each hand and each foot. "'The seventh Nummo,' said the old man, 'put the cowries in the place of the hands because men count with their fingers. He put eight for each hand, because when men first began trading they counted in eights'" (page 200). "'Beyond that number men counted by as many times eight as there were nails on the two hands, that is by eighties. Eight times eighty, or 640, was the limit'" (page 53). Griaule added that later the Dogon counted in units of eighty, called "a hundred" in French, but ten eighties were combined to form 800, the next unit. (See the chapter on currency in this book on the subject of the "Bambara hundred." The Bambara live close to the Dogon, south of the Niger River bend.)

The birth of twins is a notable event among the Dogon, and the number eight is associated closely with the rituals celebrating the birth and growth of the fortunate pair. Indeed, "eight" is omnipresent in Dogon life and belief—a spiral of eight turns of red copper surrounds the sun, and eight pillars support the traditional home, built around a square room, eight cubits on a side, the same dimensions as agricultural plots.

In the Dogon cosmology it is the spirit of the seventh of the eight ancestors who is the chief architect of the world order. This spirit wove a cloth, thereby imparting the Word to all men. The word for woven material, *soy*, also means "It is the spoken word." The word for seven is *soy*, too, for the spirit was seventh in the line of ancestors.

In many parts of the world the number seven is endowed with mystical qualities surpassing those of all other numbers. The significance attached to "seven" by the Mesopotamians affected their views of the earth and the heavens, and remains with us in the form of the seven-day week.

As late as the nineteenth century, the philosopher Hegel criticized the astronomers for not paying attention to philosophy, which had proved that there could be no more than seven planets, and for thus wasting their time looking for what could not be found!

Seven is a particularly ominous number among many African peoples. To the Kolokuma Ijo, seven is a number to avoid because of its association with the great divinities. In a later chapter I shall discuss the special significance of seven to the Kikuyu of East Africa.

The desire to avoid speaking the name of the number seven may have been responsible for the compound names found in some languages. The Mandyak of the western Sudan compound the names for nine, as well as seven: $7 = 6 + 1$ and $9 = 8 + 1$. In the Ga language $7 = 6 + 1$ and $8 = 6 + 2$.

The strangest example I have found of a compound name for a number occurs in the Mbundu language of Angola. The name for seven literally means "six-two"! I dicussed the matter with Mr. Lawrence Henderson, who had spent many years as a missionary in Angola. He told me that he had been told by an Mbundu-speaking person, who in turn had been told by an older person, one who had had no contact with anthropologists, that the original word for seven was subject to a taboo. Therefore a word for eight had slipped into its place. The designation for eight was a typically Bantu word, meaning "four-four."

A similar construction occurs in Malinke, where "six" is *woro*, "two" is *fila*, and "seven" is *woro-fila*. Delafosse states that a proper word for seven may not be spoken because it is a forbidden number; therefore an expression meaning "six-two" is substituted.

A taboo number may be handled in another manner. The speaker merely makes the gesture for the forbidden number, while the listener says the word. In that way the danger is divided between them.

The ancient Hebrews also made substitutions to circumvent a taboo. The Hebrew number symbols resembled the Greek in that the letters of the alphabet represented the numerals. Fifteen normally would be represented by *yod*, the letter for ten followed by *heh*, the letter for five. However, this combination, as well as that for ten plus six, would spell the forbidden name of Yahweh (Jehovah). Therefore the combination $9 + 6$ was substituted for $10 + 5$ and sixteen was represented as $9 + 7$ in written Hebrew. For subsequent numerals the appropriate digit was added to ten.

And how many buildings in the United States skip "thirteen" in numbering the floors? When I was negotiating the purchase of an apartment in a cooperatively owned building, I was told by the agent that every apartment in the I-line (the one with the good view) had been sold with the exception of Apartment 13-I. Nothing loathe, I agreed to purchase 13-I. The agent solemnly shook my hand, congratulating me upon my courage. Shortly after we had moved in, a young cooperator wrote an essay for our cooperative newspaper."It is a simple and accurate procedure for youngsters to learn how to count from the numbers of the floors [in our building]. This is due to the inclusion of a thirteenth floor, missing in the Empire State and other notable buildings."

SECTION 3
NUMBERS IN DAILY LIFE

Early in the twentieth century the British investigator, W. S. Routledge, described the excellent suspension bridges built of tree trunks and tough creepers by the Kikuyu of Kenya. He remarked that their counterparts in England could offer no improvement on the engineering. Intuitively the Kikuyu must have used mathematical concepts in order to achieve a stable structure suited for the transportation of man and beast.

Number entered into the daily lives of Africans in many ways—in the division of time, in the use of currency, in weighing and measuring commodities in the market place. We shall consider some of these applications.

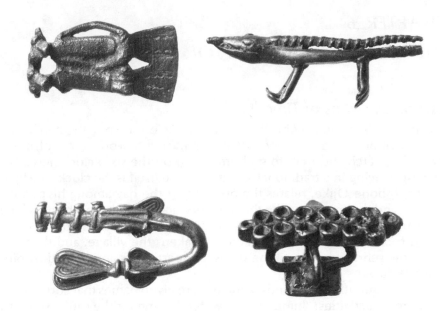

Figure 6–1 Asante brass weights for measuring gold dust. In the past every man regularly involved in trade either had his own set of about twelve weights, or had access to a set. The set of weights in the royal treasury covered a wide range of values. These weights were used in all transactions, down to the value of a penny. Historically they go back at least to the year 1600. These are a few of the thousands of existing patterns. The double spiral motif in the upper left is the sign of creation by the supreme being. In the lower right is a tiny board, about four centimeters long, for the African game of wari. The realistic peanut, the sunfish (lower left), fantastic animals and birds, and people performing their daily tasks—all were represented in these beautifully fashioned weights. Smithsonian Institution, Washington, D.C.

CHAPTER 6
THE AFRICAN CONCEPT OF TIME

Natural Divisions of Time

Few people living in a highly industrialized society are not affected by the regimentation of the clock—catch the train, arrive at work or school, break for lunch, catch the train to go home, turn on the six o'clock news. For an African living in a traditional society, Nature itself is the clock. The Igbo author, Mbone Ojike, relates the ordering of the day among his people. The sun and the moon are the timekeepers, accompanied by the quacking of partridges, the hooting of owls, the song of cuckoos, and the crowing of the roosters. The second cockcrow awakens the village, and the third cockcrow sends workers to the fields. Suppertime is announced by bird songs at dusk.

Frequently the periods of the day are given names that correspond to the main activities: "time to draw water," "time for the cattle to return home." At night the cry of the cricket provides the background to sleep, and when the cricket becomes silent, day is approaching. Farmers and herders also recognize the positions of the planets, particularly that of Venus, the morning or evening star.

The Rev. Dr. John S. Mbiti, chairman of the Department of Religion at Makerere University in Uganda, discusses the African divisions of time in *African Religions and Philosophy* (pages 20–21):

> *The day* in traditional life is reckoned according to its significant events. For example, among the Ankore of Uganda, cattle are at the heart of the people. Therefore the day is reckoned in reference to events pertaining to cattle. . . .
>
> *The month.* Lunar rather than numerical months are recognized, because of the event of the moon's changes. In the life of the people, certain events are associated with particular months, so that the months are named according to either the most important event or the prevailing weather conditions. For example, there is the 'hot' month, the month of the first rains, the weeding month, the beans harvest month, the hunting month, etc. It does not matter whether the 'hunting month' lasts 25 or 35 days: the event of hunting is what matters much more than the mathematical length of the month. . . .
>
> *The year* is likewise composed of events, but of a wider scale than those which compose either the day or the month. Where the community is agricultural, it is the seasonal activities that compose an agricultural year. Near the equator, for example, people would rec-

ognize two rain seasons and two dry seasons. When the number of season-periods is completed, then the year is also completed, since it is these four major seasons that make up an entire year. The actual number of days is irrelevant, since a year is not reckoned in terms of mathematical days but in terms of events. Therefore one year might have 350 days while another year has 390 days. The years may, and often do, differ in their length according to days, but not in their seasons and other regular events.

Since the years differ in mathematical length, numerical calendars are both impossible and meaningless in traditional life. Outside the reckoning of the year, African time concept is silent and indifferent. People expect the years to come and go, in an endless rhythm like that of day and night, and like the waning and waxing of the moon. They expect the events of the rain season, planting, harvesting, dry season, rain reason again, planting again, and so on to continue for ever.

Dr. Mbiti summarizes the African concept of time (page 19):

When Africans reckon time, it is for a concrete and specific purpose, in connection with events but not just for the sake of mathematics. Since time is a composition of events, people cannot and do not reckon it in vacuum. Numerical calendars, with one or two possible exceptions, do not exist in African traditional societies as far as I know. If such calendars exist, they are likely to be of a short duration, stretching back perhaps a few decades, but certainly not into the realm of centuries.

Instead of numerical calendars there are what one would call *phenomenon calendars*, in which the events or phenomena which constitute time are reckoned or considered in their relation with one another and as they take place, i.e. as they constitute time. For example, an expectant mother counts the lunar months of her pregnancy; a traveller counts the number of days it takes him to walk (in former years) from one part of the country to another. The day, the month, the year, one's life time or human history, are all divided up or reckoned according to their specific events, for it is these that make them meaningful.

For example, the rising of the sun is an event which is recognized by the whole community. It does not matter, therefore, whether the sun rises at 5 a.m. or 7 a.m., so long as it rises. When a person says that he will meet another at sunrise, it does not matter whether the meeting takes place at 5 a.m. or 7 a.m., so long as it is during the general period of sunrise. Likewise, it does not matter whether people go to bed at 9 p.m. or at 12 midnight: the important thing is the event of going to bed, and it is immaterial whether in one night this takes place at 10 p.m. while in another it is at midnight. For the people concerned, time is meaningful at the point of the event and not at the mathematical moment.

In western or technological society, time is a commodity which must be utilized, sold and bought; but in traditional African life, time has to be created or produced. Man is not a slave of time; instead, he 'makes' as much time as he wants.

In his novel *Arrow of God*, Chinua Achebe relates the story of the Igbo priest who "remade" time to suit his own purpose. For the Igbo people of Nigeria the yam is the staff of life. In the ritual of the annual Yam Festival celebrating the harvest, thirteen sacred yams are selected, to be kept in the village priest's storeroom for the rites of the new moon. On the evening in which each new moon first appears, the priest strikes the gong for all the people to hear; then he bakes one of the sacred yams for his evening's supper. No one dares to disturb him during this sacred ceremony.

Achebe tells how the ritual was interrupted, with dire consequences to the whole community. For it was the duty of the priest to announce the time for the harvest of the new yams when just one sacred yam remained in his storeroom. The whole agricultural cycle, the survival of the villagers, depended upon this lunar calendar. But the priest Ezeulu had been forced to take an unprecedented action; he had been summoned by the local British administrator to come to the district office, and there had been asked to assume the post of "chief," the local functionary through whom the British carried out their policy of indirect rule. Ezeulu knew the British game, and refused to play. For this act of resistance he was imprisoned. Two new moons appeared, waxed bright, and disappeared, before Ezeulu was released to return to his village.

Soon the yams were ripe and heavy, but Ezeulu refused to set the harvest day. The people were faint with hunger. In desperation a delegation of the village council pleaded with him. Ezeulu was adamant. "You all know what our custom is. I only call a new festival when there is only one yam left. Today I have three yams and so I know the time has not come" (page 259). A member of the delegation explained to the villagers: "What we told him was to go and eat the yams and we would take the consequences. But he would not do it. Why? Because the six villages allowed the white man to take him away. He has been trying to see how he could punish Umuaro and now he has the chance."

Man-Made Divisions of Time

In the urbanized areas of the continent, particularly in West Africa, the market economy is linked to a calendrical week of three, four, five, six, seven or eight days. The Yoruba, Igbo and Bini of southern Nigeria have a four-day week, as do some peoples of the Congo. In some parts of West Africa, the names of the days of the week are actually the names of the main towns in the area in which markets are held on those days. Among the Igbo people, the village cluster is divided so that the various sections have their rest periods on different days within the four-day week, or within a period of two weeks, called the "big week." On the rest day, the villagers abstain from farm work to do household chores, practice their hobbies, and visit their friends. It is on these special days that festivals, ceremonials and markets are held, frequently all in the same location. Most Igbo markets now adhere to an eight-day schedule.

Further north in Nigeria, the Tiv people have a five-day market cycle, and consequently a five-day week. Each village sees itself as the center of one of five markets, the other four meeting in different villages on the other days of the week on a rotating basis. The market names are the only names by which the days of the week are known. In addition there are large area markets held every tenth day.

The seven-day week is generally used in those areas where Islam or Christianity has gained a strong foothold. In some parts of Ghana, remote from the influence of these religions, the seven-day week is also traditional.

Age Grades

A unique method of recording time is provided by the system of age grades, common to many, but not all, African societies. Where it exists, age-grading is the most distinctive feature of social organization. To take as an example an eastern Igbo village, children born within a three-year period are organized into an age-set society to which they belong for life. Every ninth year the three sets consisting of men between the ages of thirty-five and forty-five years are reconstituted as an age grade, with police functions and responsibility for executing the decisions of the village elders. They are then promoted to the status of elders, and function as decision-makers for the village and the district.

In Achebe's book *Arrow of God*, a meeting was called to deal with the problem of the conscription of young men to build roads for the British colonial government. The village council had invited a villager who had worked for the British officer to take part in the discussion. Some council members objected to having a man of another age group joining in their deliberations. Others pointed out that this was a special meeting, and they should include the only man who knew anything about the white man.

Frequently past events are dated according to the stage of a particular age set. In his book *Facing Mount Kenya*, Kenyatta tells how each age group is named according to some important event—a famine, or the introduction of syphilis by the Europeans!

Loyalty to one's age set is equalled only by ties to one's family. Armies have been recruited on this basis, whether for the sudden but brief Maasai cattle raid, or for the conquering forces of the Zulu King Shaka in the nineteenth century. The guerilla warfare against the Portuguese in the colony of Guinea-Bissau is well organized and well disciplined—it is based upon the age-grade system of the Balante people.

Oral Traditions

Oral traditions play a great part in the dating of past events in preliterate societies. In the evenings the old people gather the children about them and recount the tales of their ancestors. On a larger scale, each royal court had its historian. It was the function of this "griot" to recite the outstanding events in the history of the reigning dynasty. The King lists, recounted orally for generations, are important means of dating events in history. Some of these lists are estimated to encompass seven or eight centuries. In fact, those of the Kuba people of the Congo region date back fifteen centuries.

Mrs. Des Forges, who has done extensive research in Rwanda, writes: "Over the longer span, time was computed according to the reign of a given monarch or his local representative in an area, or according to the life-span of some relative. Oral history still thrives in Rwanda and I found many young people who could tell their ancestors back ten generations, and one who could go back fourteen generations."

The question of trade and currency concerns us mainly in relation to the development of number systems and systems of weights and measures. There is ample evidence that from earliest times, the exchange of commodities took place both among the Africans themselves and with the outside world. The great period of the long-distance trade began with the ascendancy of Islam after 900 A.D., and we rely chiefly upon Arabic written records for our knowledge. This commerce linked the areas north and south of the Sahara. It extended from Egypt westward across the Sudan and southward down the east coast.

African currency ranged from cattle to gold, from salt to slaves. In some societies barter was the basis for trade, while in others the currency became standardized in the form of certain weights of gold. The materials used as currency usually had an intrinsic use value. Cowrie shells and beads were used in decoration; copper ingots and iron were needed for the manufacture of tools.

Barter

Before the widespread use of currency and even after its introduction, trade was carried on by barter. Over two thousand years ago Herodotus described the silent barter between the Carthaginians and the African peoples on the west coast, and this practice was still in use in recent times among peoples who do not share a common language. "Dumb barter" was a characteristic mode of trade by which ancient Ghana acquired gold in exchange for salt, copper, cloth, and cowries. The gold on which Ghana became wealthy was mined by the Wangara, who lived outside of its political control. As the Arab traveler Masudi described the trade in 950 A.D. (Shinnie, pages 45–46): "When the merchants reach this (agreed upon) boundary, they place their wares and cloth on the ground and then depart, and so the people of the Sudan come bearing gold which they leave beside the merchandise and then depart. The owners of the merchandise then return and add to the price until the bargain is concluded." In 1832 a European described an old Igbo woman silently carrying on a trade in yams. Not a word was spoken by either side. Honesty is taken for granted in such transactions.

On the local level, a spear may be exchanged for a basket of yams at the weekly market, and no currency of any kind enters the picture. As late as 1903 the Royal Niger Company carried out most of its transactions on the basis of barter; the English company made a handsome "bonus" profit by this method.

There are areas in Africa even today where the inhabitants have never paid taxes, visited a large market, or had any use whatsoever for money. In the Muyenzi region, near Lake Nyanza in Tanzania, the local daily wage in 1965 was two pounds of beans, which the worker could then exchange for other items. Suppose he wants yams. He goes to the man who has yams and needs beans, and says, "I have beans and I want yams." Then ensues a discussion on the number of yams for a measure of beans. If they cannot agree, each searches further until he can make his bargain. Some people are willing to trade food only for salt, the traditional form of currency in that area.

The Gold Trade

In the year 951 the Arab traveler Ibn Hawkal described the king of Ghana as the wealthiest in the world and claimed to have seen a receipt for 42,000 gold dinars in trans-Saharan trade. A century later Al-Bakri described the gold currency of ancient Ghana as the most valued in the entire system of trans-African trade, which was centered in Cairo. The rulers realized that they must control the supply if they were to maintain its value. Consequently, all gold nuggets belonged to the king, but the gold dust belonged to the people.

The most fabulous king of ancient Mali was Mansa Musa. In 1324 he undertook a pilgrimage to Mecca which was to have repercussions on the economy of the Middle East for years afterwards. He traveled with a caravan of sixty thousand men, and brought with him 250,000 mithqals of gold (over a million pounds sterling). He was so generous in distributing this fortune that he depressed the value of the local dinar in Egypt for twelve years.

At this time African gold mines were supplying the basis for Europe's currency. The empire of Songhai sent its gold northward across the Sahara, along with ivory, ebony, and slaves. In exchange, the empire received manufactured goods of copper and iron, brassware, European sword blades, textiles, and most important of all, salt.

The Asante system of evaluating gold dust by weight was widely recognized in West Africa. Traders became familiar with the variety of brass figurines and geometric forms owned by Asante chiefs and merchants. They were used on jewelers' scales to balance quantities of gold dust. As an example of gold weight equivalents in the eighteenth century, 24 *damba* or 12 *taku* gold made one gold ackie, or one-sixteenth of an ounce.

On the East Coast, ivory and gold were the most desired exports. To the Swahili city-states such as Kilwa, African traders brought gold dust from the mines of Monomotapa and ivory tusks from the inland regions to be exchanged for cloth and beads from abroad. From Arab and Portuguese records we know that the rulers of Kilwa demanded certain specified weights in gold from merchants bearing cloth, ivory and gold for export.

Figure 7–1 Asante brass weights for measuring gold dust. Representational weights such as these usually have associated proverbs. There is no relationship between the pattern of the weight and the amount of gold dust involved in the transaction. M. D. McLeod writes: "An absolute scale of values existed in theory and was formerly generally understood, but each value had no connected equivalent in a specific weight design." ("Goldweights of Asante," African Arts, UCLA, V #1 (Autumn, 1971), pp. 8–15.) Smithsonian Institution, Washington, D.C.

The Portuguese, from the sixteenth century onward, attempted to take over this trade, and to discover the location of the gold and silver mines, the source of the fabulous wealth of the kingdoms of the Zambezi and Monomotapa. Although the rulers of these lands were able to conceal the location of the mines from the Portuguese for a time, they were unable to prevent the traffic in slaves that eventually destroyed their culture.

Shell Money

The most popular currency within Africa was the cowrie shell. Pictures of cowrie shells drawn by Paleolithic man appear on cave walls. The ancient Egyptians considered the cowrie a magic agent, a talisman of fertility, and in some cases used it as currency in foreign exchange. Archeologists have found millions of them in the tombs of the Pharoahs. By the thirteenth century Africa had been flooded with these small shells.

Cowrie shells were introduced from the Maldives of the Indian Ocean by Arab caravans which brought them from Egypt across the Sahara to the western Sudan region. A later method of entry was by way of the Guinea Coast ports of West Africa, with the Dutch and English acting as middlemen. This currency was treated with disdain by Europeans, partly because of the debasement of their value during the nineteenth century. An anonymous Dutch traveler wrote in 1747: ". . . those who are pleased to show a contempt of them don't reflect that shells are as fit for a common standard of pecuniary value as either gold or silver." The British trader, Hugh Clapperton, considered cowrie currency very convenient in that there was no possibility of forgery.

To European traders it seemed strange that Africans frequently preferred cowrie shells to gold coin. We are accustomed to think of money as embodying at least three functions: it is a measure of value, a medium of exchange, and a means of storing wealth. We take for granted that coins and paper notes have no inherently functional value. The traditional African attitude was quite different, and the African felt, for example, not that yams were becoming more expensive, but that cowries were getting cheaper. The European merchants had to meet the demand for currency with objects that had intrinsic value. According to a report made in the year 1592, the townspeople of Loanda (in Angola) rejected gold coins in favor of cowries in payment for slaves. In fact they were willing to accept just one *lufuku* (a numerical measure) of shells for every two in goods, a fifty percent discount. Cowries were used in a thousand different ways as decorations. They were necessary currency for such special purposes as bridewealth and payment of fines, they could be used for purchasing small necessities— but of what use were gold coins?

Even in those areas where gold was the favored currency for international exchange, cowries were in demand for small purchases—a bite of food, a clay pot. Leo Africanus wrote in the early sixteenth century: "The coin of Timbuktu is of gold without any stamp or superscription, but in matters of small value they use certain shells brought hither out of the kingdom of Persia."

Earlier Ibn Battuta, the traveler from Tangier, had reported in the mid-fourteenth century: "The buying and selling of its (Gao's) inhabitants is done with cowrie shells, and the same is the case at Malli." The writer continues with some cowrie-gold equivalents. One gold dinar (an Arabic coin) was worth 1150 cowries in Mali and Gao, while 1200 shells could be exchanged for a large mithqal, an Arabic weight equal to seventy-five grains.

Counting Cowries

An analysis of the relationship of cowrie currency to the numeration systems of various peoples is a fascinating study. Some systems of reckoning were expanded because of the demands of cowrie counting; often a special system of numeration was used just for cowrie shell arithmetic.

Dr. Raum writes in *Arithmetic in Africa* (page 58): "The economic activities of a native people have been very thoroughly studied in the case of the Ewe and, though some of the customs referred to here may have disappeared, they show particularly well the sort of multiplicative processes still known to Africans." On the Guinea Coast forty cowrie shells were strung as a unit, called *hoka* by the Ewe. The price of a commodity had to

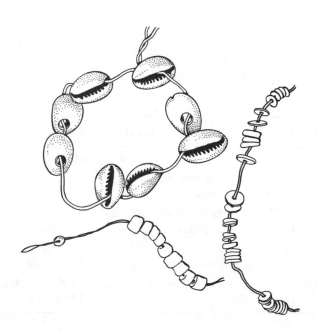

Figure 7–2 Strings of currency objects—cowrie shells, trade beads, and ivory discs.

be decomposed into factors of forty, plus a remainder. Further inland, the *hoka* consisted of just thirty-five cowries, and for this unit the shells were counted out in seven groups of five. "In quick counting," writes Raum, "the Ewe removed twenty times three cowries and added ten which gives $(20 \times 3) + 10 = 70$, or two *hoka*." When objects other than cowries had to be counted, the grouping was in pairs up to twenty (ten twos), called *amenu*. Five *amenu* formed a *ga*, or a hundred.

The Igbo people had a unique system of cowrie equivalents, with a special nomenclature. First they were counted out by sixes, and then ten groups of six were combined to form piles of sixty. Mr. Dike recalls that cowries were the common currency in his locality just a generation ago, and were carted about in bags. A few of the equivalents were:

6 cowries = *isiego* (*isii* = 6, *ego* = money)
12 cowries = *ego nabo* (2 units of money)
18 cowries = *ego nato* (3 units of money)
60 cowries = *ego neri* (10 units of money) = *ughu* (a new unit)

The next units were 1200 (20×60) and 24,000 (20×1200). The complete system is described by Jeffreys in "The Cowrie Shell." He lists number words up to 96,000,000 ($10 \times 60 \times 20^4$).

No one has yet solved the mystery of the various systems employed in counting cowries as one traveled up the Niger River from the Guinea Coast past the Niger bend, the site of Timbuktu and Gao. According to

Ounces of Gold	Name of Gold Unit	Cowrie Shell Equivalent
$3/20 = 0.15$ (approximate)	large *mithqal* (75 grains)	$12 \times 100 = 1200$ cowries 12 Muslim hundreds
$12/100 = 0.12$ (approximate)	small *mithqal* or *gros* (60 grains)	$12 \times 80 = 960$ cowries 12 Bambara hundreds
	soa (basic weight of Asante system)	$12 \times 60 = 720$ 12 Mandingo hundreds
$1/16 = 0.06^+$	ackie (coastal)	$12 \times 40 = 480$ 12 strings

Table 2.   *Cowrie and Gold Currency Equivalents about the Year 1500.*

(Based on Marion Johnson: "The Cowrie Currency of West Africa," Part II, JOURNAL OF AFRICAN HISTORY, 1970 No. 3, page 333.)

available sources, this was the cowrie currency situation about the year 1500 (Table 2).

We have more information from Europeans of the nineteenth century. The cowries were generally counted in groups of five, and then piled into "hundreds." But a "hundred" might be 60, 80 or 100, according to the local convention. Although the Bambara called 8 × 10 cowries *keme*, they also used *keme* to identify the number 10 × 10 when counting objects other than cowries. In 1866 the French officer, Mage, gave the following equivalents and the description of the procedure (quoted in M. Johnson, Part I, pp. 38–39):

$$8 \times 10 = \text{hundred (80)}$$
$$10 \times \text{hundred} = \text{thousand (800)}$$
$$10 \times \text{thousand} = \text{ten thousand (8000)}$$
$$8 \times \text{ten thousand} = \text{one hundred thousand (64,000)}$$

The procedure was to "count by five cowries at a time, which they pick up with dexterity and speed only acquired by long practise; when, taking them in this way, they have counted sixteen times five, they make a pile of them, that is 100. When they have five of these piles they put them together, make five more, join them all together, and that is 1000. Traders and women, to avoid mistakes, usually begin by making a lot of little piles of five cowries and put them together in eights to make a half-hundred."

How can one explain these variable groupings? Some writers claimed that the base of the system was originally six or eight, as the case may be, and that the Bambara (base eight) system took over, with the modification that the higher rank was based on 10 × 8 in place of the earlier 8 × 8. But why this system for cowrie counting only? As we have seen, the Bambara used 10 × 10 = 100 for non-cowrie calculations.

A more convincing explanation, offered by both Einzig and Johnson, is that the profit is built into the numeration system. For example, a large-scale dealer purchases five bars of salt for one hundred thousand cowries (actually 64,000). Then he breaks the load into small quantities, so that he receives a total of 20,000 cowries for each bar, making a profit of 7200 cowries per bar over his initial investment of 12,800 cowries.

Now let us return to the cowrie-gold equivalents of the year 1500. Marion Johnson offers the logical explanation of built-in profit to account for the variations. An *ackie* (480 cowries) of goods at the coast sold for a *soa* (720 cowries), the basic weight of the Asante system, at the first great market to the north as one traveled up the Niger River. At the next market the price was a *gros* (960 cowries), and beyond the Niger bend it was a large mithqal. Goods flowed north as gold flowed south, and the profits were built into the currency system.

Depreciation of Cowrie Currency

Europeans were able to manipulate the value of the cowries by bringing tremendous quantities of the shells into Africa at small cost to themselves. Beginning in the sixteenth century, there was a steady large scale devaluation, to the advantage of the European traders. As the value declined, the price of commodities in cowrie shells rose proportionally, and the numeration systems expanded to meet the demands for larger and larger numbers.

As cowries depreciated in value in comparison with coinage, it became too expensive to transport them from one market to another. Prices might still be quoted in terms of cowries, but the actual transactions for high-valued items were carried out in gold dust, kola nuts, salt, and even livestock or slaves. Within the markets, literate merchants kept written accounts in cowrie units, and settled their accounts later in terms of some more convenient medium of exchange.

Unit	Number of Cowries	British Equivalent
Tocky (or toque), a string	40	
Gallina (Portuguese galinha, 'hen')	200	6d.
Ackey	1000	2s. 6d.
Cabess (Portuguese cabeça, 'head')	4000	10s. 0d.
Ounce "trade"	16,000	40s. 0d.
(16,000 cowries weighed about 40 pounds)		

Table 3. Cowrie units and British equivalents on the Guinea Coast in the late eighteenth century.

(From Archibald Dalzel: The History of Dahomey. London, 1793.)

By the 1860s a large unit of 20,000 cowries had come into use. In Bambara country, on the Niger River, it was known by the French name captif (slave), since 20,000 cowrie shells weighed from fifty to one hundred pounds, approximately the load that one man could carry. A captif, however, did not equal the actual price of a slave, which at that time was between 4000 and 40,000 cowries.

Delafosse described the situation in the early twentieth century, after the French had established the cowrie equivalent at one thousand to the franc in their colonies. He asserted (according to M. Johnson, Part I, page 40) that in any single country the price of cowries might vary according to the supply of money and would fluctuate between the Malinke (Mandingo) thousand (actually 600) and the Muslim rate of 1000 to the franc. Some Africans made a practice of cornering the market on cowries to raise the exchange rate.

Other Currencies

Besides cowrie shells and gold, a variety of other articles have been used as currency. Archaeologists digging near the Zambezi River have found much evidence, dating back to the seventh and eighth centuries, of a lively trade in the area—gold and copper jewelry, welded iron gongs, cloth, beads,

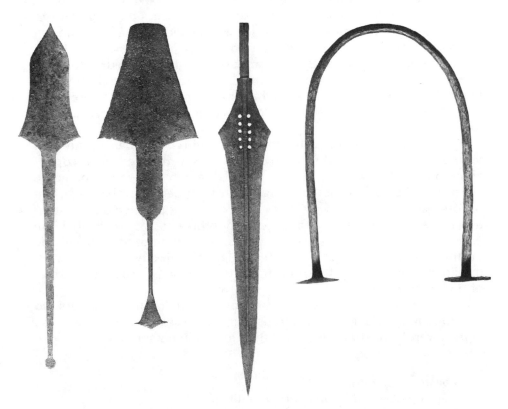

Figure 7–3 Iron objects, some several feet in length, used as currency in the Congo region. British Museum.

and huge copper crosses all used as money. A tremendous grave site of the eighth and ninth centuries, found in the copper-bearing Katanga region of the Congo, yielded pottery, copper jewelry, imported beads, and again the copper Handa crosses. The Bushongo people of the Congo region are so named because of the throwing-knife, or *shongo*, which they used as currency.

Salt, a necessary and scarce part of the diet in the African tropics, was a valued commodity in many areas. In bars of standard sizes it was used as a medium of exchange. Where kola grew, kola nuts circulated as money and were counted out in units similar to those of cowrie shells. In East Africa, cattle and other livestock almost exclusively constituted the measure of wealth.

In the early fifteenth century the Egyptian historian Al-Maqrizi wrote about Kanem (now northeast Nigeria): "As regards money, they use a kind of cloth which they make and which is called 'Wendy.' Each piece is ten cubits long, but for facility of exchange it is cut up into pieces of a quarter of a cubit or smaller. Other substances such as shells of different kinds and pieces of copper or gold are equally used in commerce and their value is estimated in an equivalent amount of cloth" (quoted in Hodgkin, page 77). The situation had not changed much four centuries later. In the kingdom of Bornu, the successor to Kanem, cowrie equivalents were stated in terms of the *gabaga*, a strip of cloth serving as the traditional currency unit.

Here cowries were counted in fours, eight cowries being equivalent to one *gabaga*. Probably the Bornu cowrie units were based on the copper coins minted in that land in the late eighteenth and early nineteenth centuries. The *rotl*, or pound (originally of copper), was the equivalent of four *gabaga*, or 32 cowries. Each time a *rotl* was counted out, an additional shell was set aside as a tally, so that three *rotl* approximated one hundred cowries. Perhaps this method was an attempt to reconcile the northern system with that of southern Nigeria, where two hundred were threaded in five strings of forty each, three strings of sixty-six, or two strings of one hundred cowries.

The market-place is the source of a variety of units applicable to currency. In the Congo area bead currency was strung in units of five; these in turn were grouped in fives, and then joined in higher units:

5 beads = one unit
5 units = 25 beads = *matanu* (literally "five")
2 *matanu* = 50 beads = *lufuku* (literally "the end" of counting)
3 *lufuku* = 150 beads = *mufuku tatu* (literally "the end three times")

In another part of the Congo region, the brass rod currency, consisting of one rod twisted around other rods to complete a bundle, gave rise to many numerical stems.

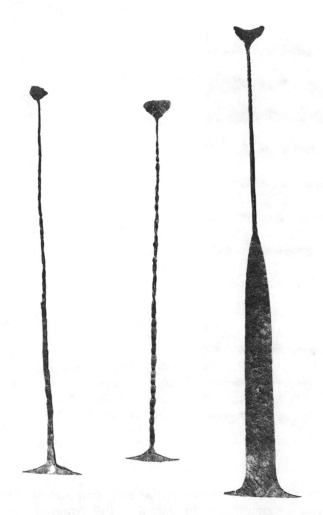

Figure 7–4   Kissi "pennies," Sierra Leone. British Museum.

In parts of Tanzania and Kenya, beads were the principal currency after the advent of European and Arab traders. For convenience they were strung in units of twenty, a bunch of ten strings forming a *fundo*. The advertisement for the Africana Holiday Village, a new tourist attraction near Dar es Salaam, reads: "The tourist will exchange his money into beads to barter for services and articles at the village" (*Daily Nation*, Nairobi, February 12, 1970).

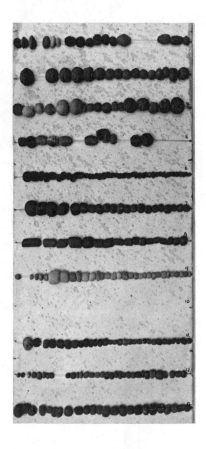

Figure 7–5 Pre-European trade beads. Most of these beads are made of glass and came from India. There are also cornelian, agate and quartz beads. National Museum of Tanzania.

Taxation

The collection of taxes involved arithmetic computation. Only in some instances can we determine the extent of this computation. In ancient Ghana, taxes were collected on every load of goods which entered or departed from the land. Al-Bakri writes: "For every donkey loaded with salt that enters the country, the king takes a duty of one gold dinar, and two dinars from every one that leaves. From a load of copper the king's due is five mithqals and from a load of other goods ten mithqals."

Revenue from the imperial domains of Songhai filled the coffers of the state apparatus. Every district had its tax collector, called *mundyo*, and

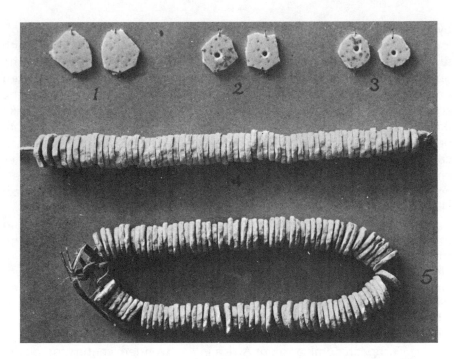

Figure 7–6 Pre-European trade beads (African). Ostrich egg shell ornaments from Mara district, northeast Lake Nyanza (Victoria), Tanzania. This display shows the stages of making the beads and their arrangement in the form of a necklace. National Museum, Tanzania. (Donated by E. C. Baker)

a minister to coordinate the whole system. Taxes were imposed on the peasants as well as on visiting traders.

The kings of Dahomey exacted tribute by adjusting the number of cowries on the string. Instead of the standard forty, there were only 34 or 37 per string. Thus the king, while paying fixed market prices, obtained his purchases at a discount of up to fifteen percent. This method of taxation may be compared to the royal gold-weights of the king of the Asante, which were heavier than those of the merchants. Thus he was guaranteed a larger quantity of gold dust for each unit.

Taxes had been known on the coast of the Indian Ocean for centuries. The Swahili chronicles mention the *zakat*, a tax on every male Muslim of full age. This income provided funds for religious and charitable expenditures and for public works. Later, a tax was imposed on the transportation of slaves.

Taxation in the Buganda kingdom in East Africa was a well-organized procedure. The country was divided into counties, each supervised by a territorial governor. Each administrator was responsible for the collection of the levies of cowrie shells, hoes, and barkcloths, which he then turned over to the prime minister. The system was based on a census of the peasants, who were taxed at a definite price per head. In return, a fixed proportion came back to the local chiefs for the administration of the counties. Commerce was strictly regulated by the authority of the Kabaka (king). Although the market places belonged to the local chiefs, a private person was permitted to open a market, and each day he was entitled to collect ten percent of the earnings of every merchant—a sizeable sales tax. Tribute then was paid to the chiefs and finally to the Kabaka, who lost no opportunity for the collection of revenue—head tax, sales tax—almost like our society.

In the relatively unorganized neighboring lands, tribute was paid to the rulers on a more casual basis. The acceptable tax payments to the royal court of Rwanda consisted of any goods useful for the sustenance of life or comfort. The tax structure was a complex one, involving different kinds and amounts of payment, depending on the region, status etc. When the bearers arrived with the goods, they paraded them before the king in a long procession. A royal official risked dismissal if the contribution from his region was deemed inadequate.

European domination of Africa in the twentieth century forced the African into the wage-money economy. The colonial governments needed labor to cultivate the lands, to build roads, to construct their buildings. But the African's economic system placed no value on pecuniary awards, and the average person could not be persuaded to leave his home for this purpose. One answer was taxation—a "hut tax" was levied on every house and had to be paid in official currency. Even today some African peoples distrust coinage, as Bohannan writes of the Tiv of Nigeria.

In many African societies it is customary—indeed, expected—that a man, after he has been married for a time, will take on a second wife, provided he has the means to support the families of both wives. Each has her own house, in which she reigns supreme, while the man also has his house. The wives take turns supplying his needs and enjoying his company.

The advent of colonial rule put a damper on this practice. After the introduction of the hut tax, only the wealthier Africans could afford to maintain the customary separate houses for their wives, or for even one wife and her children.

Soon after the introduction of the hut tax, a personal tax was levied, to be paid in official government currency. Women sold the produce of their small plots, after setting aside the barest minimum for their families. People unaccustomed to using currency were forced to seek employment on white-owned farms to earn the necessary cash.

We have remarked on the fear of Africans to count the number of people or houses in their villages. The traditional superstition about counting living creatures or valued possessions has been largely superseded by fear of the tax collector! The African, when asked how many people or houses there were in his village, would naturally give a lower figure than the true one. Sometimes it worked to their disadvantage. A Yoruba friend told me of a government survey to determine the population of each village, and the natural tendency on the part of the villagers to understate. Soon word got around that benefits were to be distributed on the basis of these figures. Immediately the population increased threefold!

Bridewealth

Although minted currency is now the rule, traditional societies still require special purpose money for the payment of bridewealth, for sacrifices, and for special occasions.

The concept of "bridewealth" has been thoroughly misunderstood by Europeans, who interpreted it to mean that African men "bought" their wives. The early missionaries were horrified by the practice and tried to abolish it. Its true function can best be described in the words of the Rev. Dr. Mbiti (page 140):

> The custom of presenting a gift to the bride's people is practised all over Africa, though in varying degrees. Different names are used to describe it, such as "bridewealth," "bride-gift," "bride-price," "dowry" (wrongly in this case) and "lobola." Most of these terms are either inadequate or misleading. The gift is in the form of cattle, money, foodstuffs and other articles. In some societies the families concerned may exchange brides. In others, the bridegroom (and his relatives) must in addition contribute labour; and in matrilocal societies the man lives with his parents-in-law working for them for some years in order to "earn" his wife.
>
> This marriage gift is an important institution in African societies. It is a token of gratitude on the part of the bridegroom's people to those of the bride, for their care over her and for allowing her to become his wife. At her home the gift "replaces" her, reminding the family that she will leave or has left and yet she is not dead. She is a valuable person not only to her family but to her husband's people. At marriage she is not stolen but is given away under mutual agreement between the two families. The gift elevates the value attached to her both as a person and as a wife. The gift legalizes her value and the marriage contract. The institution of this practice is the most concrete symbol of the marriage covenant and security. Under no circumstances is this custom a form of "payment," as outsiders have so

often mistakenly said. African words for the practice of giving the marriage gift are, in most cases, different from words used in buying or selling something in the market place. Furthermore, it is not only the man and his people who give: the girl's people also give gifts in return, even if these may be materially smaller than those of the man. The two families are involved in a relationship which, among other things, demands an exchange of material and other gifts. This continues even long after the girl is married and has her own children. In some societies if the marriage breaks down completely and there is divorce, the husband may get back some of the gifts he had given to the wife's people; but in other societies, nothing is returned to him.

Figure 7–7   The iron double hoe is an essential part of the traditional bridewealth in some areas.

In *Things Fall Apart* Achebe describes the delicate negotiations to determine the amount of bridewealth to be paid by an Igbo suitor. After a few drinks and small talk about every subject but the purpose of the visit, Ukegbu, the suitor's father, finally announced why he had come. Then

Obierika, the father of the prospective bride, handed him a small bundle of sticks. Ukegbu counted them—there were thirty. He turned to his brother and his son, and suggested they go outside to whisper together. When they returned, Ukegbu passed the bundle of sticks back to Obierika, but now there were only fifteen. After some discussion his brother added ten sticks, and returned the bundle to Ukegbu. Finally they settled upon twenty bags of cowries as the bridewealth.

After the guests had left, the bride's family discussed customs in other villages. In some villages they haggled and bargained as though they were buying a goat in the market. In another, the suitor just continued to bring in bags of cowries until the in-laws told him to stop. This custom always led to a quarrel. Of course, they concluded that their way was the best.

With the disappearance of special purpose money in recent times and the introduction of general currency, brides "entered the market." In parts of Nigeria it was necessary to set legal limits on the amount of cash that could be demanded, based on such factors as the girl's education. When I mentioned this to a Tanzanian friend, he remarked: "We don't care about her education. We just want to know that she's a virgin!"

The conflict between the traditional and the "new" is most pronounced on this subject of bridewealth. In rural areas the customary cattle, goats, cloth, etc. are expected in exchange for the bride. But the young men of the cities will have nothing to do with these old customs.

In Zambia today the customary *lobola* varies from five pounds to over one hundred pounds, and is frequently influenced by the girl's education. A young man wrote to the Zambian counterpart of the "Advice to the Lovelorn" column of the newspaper, to complain that his sweetheart's parents demanded two head of cattle plus five pounds in cash. He was a miner, and did not have access to cattle, but the parents would hear of no compromise (from *Tell Me Josephine*, ed. by Barbara Hall).

Another heartsick youth related that the family of his beloved demanded cattle as *lobola*, while his people did not "own cattle for this sort of thing. Easier for me to deliver wild buffaloes from the forest. . . . But at the thought of losing her, tears come to my eyes." Josephine replied: "Few girls are worth a dowry of buffaloes. Why not save your money and go to buy cattle from the tribes that keep them? It is wise to respect tribal custom if possible."

Josephine is flexible. To the young man who left his tribal home as a child, she gives this advice: "Since you are paying the lobola I think you have the right to choose the wife. . . . Stand firm—marry the girl of your choice. Why let old customs spoil your life? Apologize to your parents and send them gifts."

CHAPTER 8
THOSE FAMILIAR WEIGHTS AND MEASURES!

Today the United States is the only major nation still using the British imperial system of weights and measures—the system based on the foot and the pound. In 1965 the British Parliament recommended that within a ten-year period the country effect a change to the metric system. This change is now taking place. No longer will the British schoolboy or girl devote a good part of his mathematical training to the confusing subject of converting yards to feet to inches, based on 3 and 12; pounds sterling to shillings to pence, based on 20 and 12; pounds to ounces, based on 16 if avoirdupois or 12 if Troy; gallons to quarts to pints to ounces, based on 4, 2 and 16. All he needs to know is how to multiply or divide by powers of 10—in short, how to shift the decimal point.

In commemoration of their participation in this great event, the governments of Kenya, Tanzania and Uganda issued a series of beautiful stamps, each illustrating a different aspect of conversion from the British imperial system to the metric. Meanwhile the United States, the leading industrial power in the world, retains the vestiges of primitive attempts at a standardization of measurement.

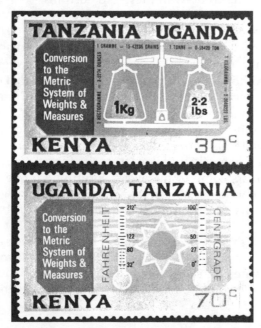

Figure 8–1 Special postage issue commemorating conversion to the metric system in Kenya, Tanzania and Uganda.

The British imperial system originated as a standardization of measures based on the parts of the human body:

Inch: from the tip to the knuckle of the thumb.

Cubit (Biblical measure): from the elbow to the end of the middle finger, or six palms.

Foot: four palms, or sixteen fingers. A thousand years ago it was standardized as 36 barleycorns "taken from the middle of the ear."

Yard: from King Edgar's nose to the tip of the middle finger of his outstretched arm.

Fathom: from tip to tip of the fingers of the outstretched arms.

Acre: the amount of land that could be plowed in a day by a yoke of oxen.

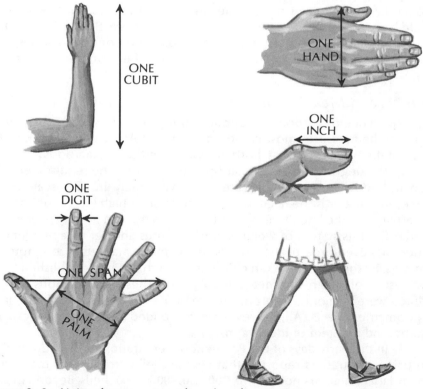

Figure 8–2   Units of measurement based on the human body.

A PACE = ONE YARD

Even in the time of Leonardo da Vinci, the arm length in Italy had not yet been standardized; each city had its standard "braccia," but it varied from Florence to Milan to Rome.

The resulting problems were many, and begged a solution. As trade between nations increased, it was necessary to arrive at standardized sets of weights and measures. A door manufactured in Florence had to be made to fit a cathedral in Milan. An international incident might result if the English merchant was accused of dishonest measurement by a French customer.

The British Magna Carta spelled out weights and measures for wine, ale, corn and cloth. Other standards were likewise imposed, and the units of the imperial system were fixed.

A decimal system of measures was first proposed early in the seventeenth century by Simon Stevin, an inspector of dikes in Holland. The metric system, based on decimal numeration, was officially launched by France in 1790 and was disseminated widely through Europe by Napoleon. Later proponents were Thomas Jefferson and John Quincy Adams, who reluctantly advised the United States in 1821 not to adopt it "because it would be hazardous to deviate from the practice of Great Britain." A century later British industrialists argued against metrication on the grounds that they did not want to deviate from the practice of the United States, their biggest customer!

Weights and Measures in Africa

The experience of Europe was duplicated in Africa. It is not surprising to find that the measures most commonly found in Africa are very similar to those of the British system—both are based on the human body.

The degree of standardization depends upon the requirements of each individual society. The greater the involvement in wide-scale commerce, the greater is the need for standardization. A high level of uniformity was attained in the late fifteenth century, when the vast empire of Songhai standardized its system of weights and measures and set up a network of inspectors to enforce the standards in the many markets throughout its domains. By this time the Akan chiefs and merchants had established their own system of weights to measure gold dust. These justly famous Asante artifacts were fashioned in bronze in the form of figurines, or of inanimate or geometric objects. All, however, corresponded to a calculated scale of weights and a system of ideographs.

In the early days of the Portuguese penetration of the Congo, five centuries ago, traders reported that cowrie shell currency was measured in standard containers holding 40, 100, 250, 400, or 500 shells. Beyond these, the measure that consisted of containers amounting to 1000 shells was a *funda*, 10,000 was a *lufuku*, while a *cofo* consisted of 20,000 cowrie shells. In contrast, the efforts of the British Niger Expedition to substitute measure by container in place of the actual counting of cowries proved to be unpopular. Perhaps the nineteenth century African, his experiences with Europeans fresh in his memory, was afraid that he would be cheated.

The fact that a government has adopted standards does not mean that the average citizen of the land restricts himself to the imperial measure when he comes to the market to exchange his surplus grain for a few hens. In fact, in Africa, standardized measures have been the exception rather than the rule. The proliferation of local measures in use until recent times is beyond listing. Among the Kpelle people of Liberia even today, only rice, of all the commodities exchanged, is measured by a system of co-ordinated and interrelated containers because of the prime importance of this grain in the economy. Most other measurements are merely approximate.

Figure 8–3 Engraved calabash bottles, Kamba, Kenya. The figures and symbols often tell a story. British Museum.

Typical of the measures used in Africa are those of the Ganda people of Uganda, where units of length refer to the human body, and units of capacity depend upon the article involved. In their language, *mukono* is a cubit, the distance from the elbow to the tip of the outstretched middle finger. They also have words for the fathom, measured from the tip of one outstretched arm to the tip of the other, and the span, from the tip of the thumb to the tip of the extended second finger. Wooden poles for home building are measured in feet by actual pacing. Dry measure depends to some extent the article to be sold. The *kibo* is a basket holding about ten pounds. Salt might be sold in small packets, or as a load of thirty to forty pounds called *lubya*. A *kiribwa* refers to a package of coffee beans weighing about twenty pounds, and *lutata* is a bundle of approximately thirty pounds of sweet potatoes. Beer is measured in gourds of specific sizes. These were the measures in common use during the period in which Uganda, as a British colony, was bound to the British imperial system of weights and measures.

A few decades ago, the Swahili, on the East African coast, were using these measures:

Length: *shibiri* = span, from the tip of the thumb to the tip of
         the little finger, about 9 inches.
         mkono = 2 *shibiri* – cubit, about one-half yard.
         pima = 4 *mkono* – fathom, about 2 yards.

Capacity: *kibaba* = pint
           kisaga = 2 *kibaba* – quart
           pishi = 2 *kisaga* – half gallon

Weight: *wakia* = ounce
         ratli = 16 *wakia* – pound
         frasila = about 36 *ratli*

In actual practice there was considerable variation from this table. One could purchase *kibaba cha tele* or *kibaba cha mfuto* of grain—a heaped up or a levelled off pint measure. The cubit of cloth might be *mkono mkamili* or *mkono mkonde*, full or short. And whose arm is used as the measure of a cubit, the buyer's or the seller's? In the old days a man would induce his long-armed friend to come to the market to shop for him. More recently a standard length for the cubit was adopted and prominently displayed by shopkeepers.

In a current Swahili language instruction book one reads:

Weight: *aunsi* = ounce
            kilo = kilo
            ratli = pound

Capacity: *pishi* = 2 litres
            kibaba = one-half litre
            debe = 4 gallons

Length: *yadi* = yard
            futi = foot
            inchi = inch
            futikamba (lit. foot-rope) = tape measure
            maili = mile

The imperial British measures have virtually displaced earlier systems.

Dr. Richard Pankhurst has made an exhaustive study of the measures, weights and values in use among the various Ethiopian peoples throughout their history. The measures of length again include the ancient cubit, the fathom, and the span, as well as smaller measures based upon the thickness of a finger, the length of a joint, and the width of four fingers together. The cubit was affected by the arm length of the one who measured. One narrator noted in the early nineteenth century that a man with a long arm measured from the elbow to the tip of the middle finger; one with a middle-sized arm doubled the cloth back over his middle finger to the middle knuckle; while a short-armed person turned back the material the whole length of the middle finger. The author did not specify who performed the classification of arm lengths. In the highlands of Eritrea the purchaser was permitted to choose any person he wished to serve as a standard of measurement. In other areas the dealer used his own arm; the trader with the longest arm had the most customers. But by the 1930s standard iron measuring rods were in circulation.

One can imagine what a complicated procedure the purchase of a piece of cloth might be. An Ethiopian proverb states: "Measure ten times, tear the cloth once."

Traditional imprecise measures were used in Ethiopia until the policy of modernization was introduced in the late nineteenth century and land registration was required in the twentieth century. Then we find that the meter was mandated as the standard unit of length. But the cubit did not disappear. It was still in use in some areas in 1967.

Reminiscent of the English acre is an Ethiopian measure representing the amount of land that could be ploughed in a day by a pair of oxen. A long distance is measured in terms of the number of days' journey by foot or by mule. This standard of measurement is found in the Bible, and throughout the world.

Measurement of capacity in Ethiopia (as elsewhere) was conceived in terms of the article to be measured, the quantity involved, and the purpose of measurement. Since honey and grain played an important role in taxation, the measures for these commodities were closer to standardization. Items such as grain, which we are accustomed to purchase by weight, are measured in containers in a market that has no weighing mechanisms. Like the old recipe that called for a pinch of salt and a handful of raisins, many measures of capacity were based on the human body—a handful, an armful, a mouthful. A larger unit, based on animal capacity, was the mule-load.

The most common measure of capacity was based on a wide variety of containers, locally made of clay, wood, skin, horn, basketware, or gourd. Later the people accepted imported containers, and it was not uncommon to find liquor bottles or cans from tinned foods used as measuring devices.

Thousands of miles from Ethiopia is the city of Benin, now part of Nigeria, but formerly a city state of great wealth. In a paper written for a course at a teachers' college, Peter Idehen describes the weights and measures used in Benin in olden times:

Comparison: Distances and heights are always compared as long or short, large or small, thick or thin, etc. They compare tall or long things to tall grown palm trees or snakes, height of an adult, etc.

Measuring: They measure short distances with their feet. They do this by counting the number of times they placed one foot in front of the other foot. They can also use the distance between the thumb and the first finger to measure very short distances, such as a piece of wood to be carved. If they wish to measure distances up to a hundred feet, they pace the distance. Very long distances, such as the distance between two towns or villages, are measured by stating it in days' journeys or by time between breakfast and lunch or dinner. The following measures are employed:

Foot = the length of a man's foot
Span = the length of the palm
Cubit = the length of the forearm
Pacing = a pace is about one yard

These and other primitive measures were employed by the Binis. There were no standard measures of length developed by the natives; and these primitive measures were used till the introduction of British standard lengths immediately after the Benin City expedition in 1897.

Measuring by Weighing: The Binis are not keen traders and cannot write, hence they have not been able to develop standard measures for weights. Foodstuffs are sold locally, either cooked or uncooked. The measure of cooked foods is simply by approximation. The weight is determined by feeling two objects of the same kind. This is done by holding them in the palms and feeling the weight; another method is to lift the object if it is large and bulky. The weight of loads was calculated as man's head-load, two men's head-loads, etc. Certain things were measured in baskets and calabashes. In fact what happened was that people fixed their own measures and others were obliged to use them. The use of such measures by others depended on the demand for such articles. The method was very unsatisfactory because the selfish people often cheated the public.

Figure 8–4 Carved calabash from Ilorin, northern Yoruba region, Nigeria. The calabash is used as a standard of measure, and is an object of beauty, with its geometric patterns. British Museum.

This condition was a bit improved when the Europeans, especially the Portuguese, came to Benin in the sixteenth century. They introduced bottles when they brought trade gin to the country. They brought kegs of gunpowder and many other containers. They introduced spoons which the Binis call *Ekuye*. (People say that *Kuye* is a Portuguese word.) The people used these containers as measures so that even up till this day one finds the Bini women using bottles in measuring liquids. They have half pint bottles and pint bottles. They also use empty gun powder kegs for dry measure; this they called *Epipa*. Gun powder is still measured in this way.

Today the situation is different. Idehen continues:

It is now a common sight to see scales in all villages and it is equally common to hear even the oldest grandmother say *Esike* for scale.

CHAPTER 9

RECORD-KEEPING: STICKS AND STRINGS

Although most African societies did not, as far as we know, have alphabetic writing, it would be incorrect to say they had no written language. Some—the Hausa, Kanuri, Fulani, Malinke, and Swahili—adapted the Arabic alphabet to their own languages, just as many Europeans adapted the Latin alphabet to their languages. Others, such as the ancient peoples of Nubia and Kush, as well as the Vai of Liberia in recent times, developed their own scripts. There is evidence of the existence of both hieroglyphic and more advanced writing in the era before the colonial invasion. Much more research needs to be done.

We know that Africans used pictographs and ideographs to send secret messages, and to mark property. Proverbs are inscribed on Asante goldweights and Igbo calabashes and love messages are incorporated into Zulu beadwork. Tally sticks and knotted strings were used for all sorts of numerical records—the passage of time, financial transactions, scores in games.

Okoye had come to his neighbor Unoka to collect a long-standing debt. But Unoka was not a man to worry about his obligations. Laughing until the tears stood in his eyes, he bade his visitor look at the far wall of his house. There, on the shiny wall of red earth, were groups of lines drawn in chalk. There were five groups, and the smallest had ten notches. Unoka explained to his guest that each group represented a debt to someone, and that each stroke meant one hundred cowries. His smallest debt was one thousand cowries. But Unoka was not about to pay off, and Okaye departed empty-handed (Achebe: *Things Fall Apart*, page 8).

A young man of the Chagga people on the slopes of Mt. Kilimanjaro is about to set out on a twelve-day journey. He leaves his wife a length of banana plant fibre in which he has tied twelve knots. His wife unties one knot each day as she anxiously awaits his return. Under similar circumstances, a Bwende man ties knots in two strings, one for himself and the other for his wife.

A pregnant woman ties a knot in a string at each full moon, to know when she is about to give birth. In Zimbabwe (Rhodesia) the father-to-be cuts a stick, scorches it in the fire to get it dry, and marks it with a notch at the beginning of each month. Indeed, the last month of pregnancy is called the "month of the staff." In olden times the stick was saved as a record of the day of the child's birth. No doubt its marks included some reference to the time he was born.

Among the Zulu, men notched sticks, while it was a woman's privilege to tie knots. Strings were often considered only temporary records, but some people deemed them safer than tally sticks.

A young man of the Sundi people, in the Congo area, is suing for the hand of his beloved. Custom dictates that he contribute certain gifts as bridewealth. After some discussion among the families of the two young people, they agree upon the marriage settlement. Payment is usually made in several installments. A representative of each family ties a knot in his string for each installment agreed upon. Furthermore, the young man makes knots to recall the more valuable gifts. When the final settlement is made, the witnesses compare the strings to make sure they tally. The knotted string records of the bridewealth exchange are hung in the most remote corner of the house and are brought out on the occasion of a divorce or a death, as evidence of past payments. The Kamba people of Kenya, on the other hand, use notched sticks to record bridewealth contracts.

Knotted strings are used for various kinds of bookkeeping: to record the apportionment of water, count money, keep commercial records, collect taxes, count loads of reeds or grass purchased for housebuilding, record the number of ivory tusks sold. In the cloth trade, knots are made at both ends of a string, one end for purchases and the other for sales. A debtor gives a knotted string to his creditor as a promissory note. If the creditor loses the string, he is out of luck, since he has lost claim to the return of his money.

In many cultures knotted strings and tally sticks are the means of recording the passage of days, weeks, months, or years. The court historian in the kingdom of Monomotapa was obliged to tie one knot in the "royal string" for each monarch. In 1929 the string had thirty-five knots, and all the rulers could be identified, including the sister of Matope, the first ruler, whose reign dated back to the middle of the fifteenth century.

The custom of counting days and objects by tying knots is common throughout the world. The Catholic rosary, the "telling of beads," is one example. A very complicated form of knot-making was used by the Inca people of Peru to record all their official transactions. The knots in the various strands, called the quipu, represented a place-value notation based on ten. Not only were these quipus used for bookkeeping—they served as well as a means of writing history, laws, and contracts.

The Songe people have a most elaborate tally stick on which a man records all the journeys of his life. Four feet long, it also serves as a walking stick.

In former times people of southern Zimbabwe (Rhodesia) entrusted the marking of the calendar to a learned person of the village. Every morning he made a notch on the tally stick, and each month he started a new record, until a year's tallies had been recorded. Special marks were made when a debt was repaid, or the time was at hand for harvesting, for threshing, or for the approach of the heavy rains. An X instead of the usual stroke might mean that a person had borrowed something, and a corresponding X was made when he repaid the loan.

The Chagga wife who untied knots as she awaited her husband's homecoming also kept records on her wooden cooking spoon of the number of blows she received from her husband. When the spoon handle had no more room, it was time to institute divorce proceedings. Chagga boys made notches in their bows to record the number of birds they brought down; longer notches for five and ten reflected their numeration system.

Tally strings with wooden or shell markers are widespread. An East Coast calendar consists of a string stretched between two posts. The thirty wooden markers are separated into groups of ten by white beads, and each day a marker is shifted from one side to the other. Although it is called a Swahili calendar, it seems more likely that it was used by the neighboring Chagga people, who did at one time have a ten-day week. The Muslim Swahili people have adhered to the seven-day week for centuries.

Figure 9–1 Calendar board, said to have been used by Arabs and Coastal Islamic - influenced people. Date unknown; acquired by the National Museum of Tanzania in 1941. The markers are divided into three groups of ten each by two small ivory rings. On the ebony base is an inlaid crescent in ivory.

Government Records

The fabulous gold wealth of the Asante kingdom belonged to the monarch, and his subordinates were required to report their transactions. The Asantehene's treasurers kept accurate accounts, balancing them every twenty days with the aid of cowrie shells to tally their sums. Similarly, everyday market calculations in many African societies are carried out with the aid of piles of pebbles or beans.

The military draft was organized in nineteenth century Buganda by a device called "King Mutesa'a roll." It was a board with parallel rows of ten holes. A white peg was inserted into each hole, and at the end of the row was a larger hole fitted with a black peg. For every ten rows there was a still larger hole for a red peg. The white, black and red pegs represented ten, one hundred and one thousand soldiers, respectively. The powers of ten were correlated with the base ten in the Ganda number system. When the king needed men for a military operation, as when Egypt and the British attempted to subjugate his land, he handed a number of pegs to his chief. This was an order to supply the corresponding number of soldiers. When the operation had been completed, the pegs were replaced in the board, after deductions for battle casualties.

In an earlier section we described the indirect census of the population based on the collection of yams or cowrie shells. This system applied also to conscription of the army. During the nineteenth century the kingdom of Dahomey was engaged in frequent warfare with its Yoruba neighbors. Both sides were supplied with European guns. In 1840 Dahomey had a standing army of 12,000, including 5000 of the famous women warriors. The government also had at its command a number of reserve units. Each draftee was represented by a pebble, and every province sent to the capital a box containing the number of pebbles equivalent to the number of soldiers on call.

Of all the African states, Dahomey kept the most accurate statistics on births and deaths, the number of people captured or killed in battle, and those who died of disease. These accounts were kept by means of boxes of pebbles. We do not know how the accuracy of their body count compares with that of our present generals!

Tally sticks have been used by non-literate peoples the world over, and also by the literate. The British Exchequer tallies continued in use from the twelfth century until 1826, as a record of taxes owed and paid. A revolutionary attempt had been made in the reign of George III to substitute pen and paper bookkeeping, but in vain. It was not until decades later that use of tally sticks was abolished. By then a considerable number of them had accumulated; how could the British government dispose of them? Charles Dickens describes the dire consequences of government red tape and inefficiency:

The sticks were housed in Westminster, and it would naturally occur to any intelligent person that nothing could be easier than to allow them to be carried away for firewood by the miserable people who lived in that neighborhood. However, they never had been useful, and official routine required that they should never be, and so the order went out that they were to be privately and confidentially burned. It came to pass that they were burned in a stove in the House of Lords. The stove, over-gorged with these preposterous sticks, set fire to the panelling; the panelling set fire to the House of Commons; the two houses were reduced to ashes; architects were called in to build others; and we are now in the second million of the cost thereof.

Figure 9–2 British tally stick, used for many centuries to record official transactions, until Parliament abolished them in 1826.

SECTION 4
MATHEMATICAL RECREATIONS

Patterns of numbers, geometric patterns, rhythmic patterns: all are part of African daily existence. This section deals with recreational activities in which patterns play a dominant role. Patterns of numbers are basic in the game of "transferring" played throughout the continent by children and adults alike. Games of chance also depend upon the pattern of the outcomes when cowries or nuts are tossed. In earlier times, scholars in Muslim Africa devoted their time to construction of magical arrays of numbers. The geometric patterns that children draw in the sand find their way into the designs of woven cloth and carved wood artifacts.

Before the introduction of formal schooling by Europeans, African children learned all the wisdom they needed to carry on life in the tradition of their ancestors from their parents, their older brothers and sisters and kin, and from the elders of the village. Games formed an integral part of the child's education; from earliest childhood his play trained him to live in his particular cultural milieu. This section includes some of the games of African children as well as some recreational activities of the adults.

This may be the appropriate place to say a few words about African music, one of the great contributions to the arts of the world. The fantastic accomplishments of the African drummer depend upon a highly developed sense of rhythmic pattern. The leader announces the theme on his drum, and the other drummers elaborate upon it, each playing a different rhythm. Africans develop the ability to hear the various rhythms, and the dancer expresses each one with a different part of his body simultaneously. The well-trained orchestras of Buganda, the xylophone combinations of southeast Africa, and the choral groups throughout the continent employ the principle of combining several individual rhythms to produce a pleasing unity.

The classical poetry of Rwanda rests upon a formal structure of three elements—rhyme, rhythm and tone. Kinyarwanda is a tonal language— different levels of pitch express different meanings of the spoken word— and the tonal and rhythmic patterns of each line of poetry follow a conventional pattern.

Figure 10–1   Game board for wari (or oware), played by the Asante of Ghana. Many other West African peoples play the universal African game on a board having two rows of six holes. The background kente cloth is woven in geometric motifs, each with its own name and special significance. Private collection of D. W. Crowe, Madison, Wisconsin.

CHAPTER 10
GAMES TO GROW ON

Counting Rhymes and Rhythms

Children in Africa, as in other parts of the world, learn finger counting rhymes even before they are aware of the number sequence. Some rhymes go only to five, while others continue as far as twenty. Most stop at ten, corresponding to the number of fingers. In some areas the rhymes are based on a twelve system or give special emphasis to multiples of three or four.

Here is a popular five-finger Swahili verse, with an element of daring, translated freely: "Let's go! — Where? — To steal! — What about the police? — I'm out of this!" The counting begins with the little finger and ends at the thumb; with each phrase a finger is ticked off.

Many rhymes have nonsense words, or words that are obsolete or of foreign origin. The children of the Taita Hills, in southeastern Kenya, sing counting rhymes with words having no obvious meaning. According to John Williamson, an African adult made the discovery that some of these words had once been in the local language, but had been obsolete for a century. Amazingly, they were preserved in the children's songs. The children sing these words to a simple tune when they play games and even use the verses in their arithmetic work. In some districts they accompany the song by bending down the fingers of the right hand with those of the left, beginning with the little finger, and then continuing with the little finger of the left hand. These gestures are absolutely unrelated to the formal system of finger counting of the Taita people.

Counting songs are among the first items of the Venda children's musical repertory. They, too, accompany the words by counting on the fingers. Using their right index finger, they first tap the little finger of the left hand, then each consecutive finger until the thumb is reached. Counting on the right hand starts with the thumb and proceeds to the little finger, each in turn being grasped by the thumb and first finger of the left hand. Sometimes the child claps his hands when he reaches ten. The actual number words embrace several languages besides Venda — Thonga, Sotho, Afrikaans, English.

Venda children use a counting song to choose a child to perform an unpleasant task — the last one is the loser. The children good-humoredly shout "Witch" at the odd child.

A special Venda song for counting legs is a jumble of several languages and nonsense syllables. The children sit in a row with their legs outstretched in front of them. The singer points at each leg in turn as he goes down the row reciting the verses and comes back in the opposite direction when he reaches the end of the row. The leg which is tapped on the last word, *mutshelwa*, "guilty one," is withdrawn, and the singer begins again with the first leg in the row. Clever children can calculate where to sit in order to be counted out satisfactorily so that they are not last, or if they are counting, how to work it so that a particular person is selected. Here is a free translation of the verses:

> This one is a child who is just beginning to stagger about,
> This one is the sound of small reed-pipes,
> The reed-pipe of Mangayengaye,
> Masulu kungwa-kungwa, the calabash.
> Carried the chief's pitch-pipe,
> And it became bewitched, it is the guilty one.

In a Shona game, played as they sit around the fire at night, the children must listen carefully and count accurately. While an older man recites certain verses in a rhythmic pattern, the children count the number of principal beats. The teacher may vary his speed, speaking rapidly, then slowly. Or he may enunciate the words very distinctly, so that the audience loses the rhythm. All this is accompanied by a great deal of laughter and clapping. The losers are subjected to some good-natured mocking at the hands of the star students.

Three in a Row

Tic-tac-toe, three in a row. Several versions of this game are found in Africa, all more complicated than the familiar "noughts and crosses." It may be played on a board, or on lines drawn in the ground. In Zimbabwe (Rhodesia) young men and boys play the game on a board. Each of the two contestants has twelve stones; Player One's stones must be distinguishable from those of Player Two. There are three separate stages in the game (Figure 10–2a).

In the first stage each player in turn places a stone at an intersection not already occupied by his opponent's stone. The object is to form a line of three counters in any direction. The player who succeeds in forming a line of three is entitled to remove any one of his opponent's stones. Once a player has formed a line of three, he may, on the next move, remove one of the stones in the line, and place it at a different intersection on his subsequent move, in order to capture another of the enemy's counters.

Stage two begins when each player has placed all twelve of his stones on the board. They then proceed to move the stones one space at a time along a line, again with the object of completing lines of three.

Obviously some of the stones will have been captured; otherwise the twenty-four stones are locked into position at the twenty-four intersection points, and no move is possible.

The third stage occurs when one of the players has only three stones left. He may then move a stone to any free intersection point on the board. When one player is reduced to two stones, he has lost the game.

The version called "African Morris" in the pamphlet "Ancient Games" (Cooperative Recreation Service) is played by the Asante, with rules similar to those in Zimbabwe. This game is indeed ancient. 3300 years ago a playing "board" was cut into a roof slab of the ancient Egyptian temple at Kurna. It has a cross in the center square.

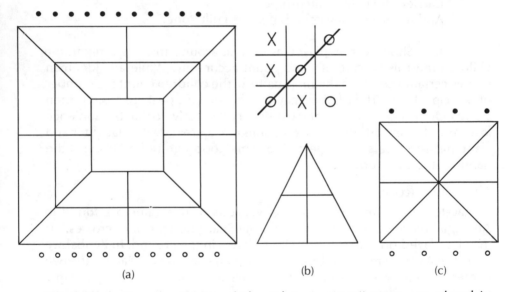

(a) (b) (c)

Figure 10–2 Several versions of the "three-in-a-row" game are played in Africa, all more complicated than the familiar "noughts and crosses." Versions (a) and (b) are popular among boys and men in Zimbabwe (Rhodesia); (c) is played by Asante children.

A similar game, The Mill, or Nine Men's Morris, is described by Geoffrey Mott-Smith in *Mathematical Puzzles*. He calls it "an ancient game that may have had a common origin with Tit-Tat-Toe (sic). . . . As a subject of mathematical inquiry, Mill is a fascinating combination of ideas. The basic three-in-a-row principle, completely exhausted in Tit-Tat-Toe, is given vitality in two ways. (a) The board is somewhat enlarged, but is still so confined that the player continually feels 'With a little thought I could analyze this game completely!' (b) A game of placement is combined with a game of movement. It is no longer sufficient to be the first to make three-in-a-row" (pages 132–134).

Another Zimbabwe version is played on a triangular network; it is called *Tsoro Yematatu* (*tatu* means "three"). There are seven intersections, and each player uses just three stones, which may be moved anywhere (Figure 10–2b). The first to complete a line of three is the winner. This is more difficult than it would seem at first glance, since there is only one free intersection available, once all the stones have been placed.

Asante schoolchildren play on a square network, marked on the ground (Figure 10–2c). Each player has four sticks or colored stones. After each player has placed his stones on the intersection points, one at a time, he may move one of his markers one space along a line. The object, again, is to place three stones in a row.

Networks

Early in this century the Belgian, Emil Torday, lived for some time among the Shongo people of the Congo area. One day he went up to a circle of small children playing with sand. "I was invited to sit down, and one of them, Minge Bengela, divested himself of his loin cloth and offered it to me as a seat. This bettered Sir Walter Raleigh's action, as my young gallant was devoid of all other clothing. The children were drawing, and I was at once asked to perform certain impossible tasks; great was their joy when the white man failed to accomplish them" (page 213).

The task was to draw each of the figures in the sand without lifting the finger or retracing any line segment (Figure 10–3).

You may have tried to draw more simple figures, or networks, without lifting your pen or retracing. Try those in Figure 10–4.

Networks a, c and d can be done, but network b is impossible. To determine whether a figure is traceable, we must examine the vertices to see how many are odd and how many are even. A vertex is a point at which two or more line segments meet. Let us analyze network a, which has five vertices. Two line segments meet at vertices A and D, three at B and C, and four at E. Vertices B and C are called odd vertices, since each is the meeting point of an odd number of line segments, and A, D, and E are called even vertices.

We can make a table of the number of line segments which meet at each vertex (Table 4):

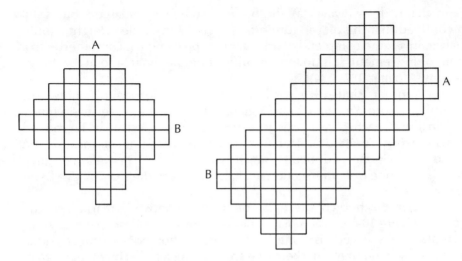

Figure 10–3   Shongo children draw these networks in the sand in a continuous line, without lifting the finger. (After Torday)

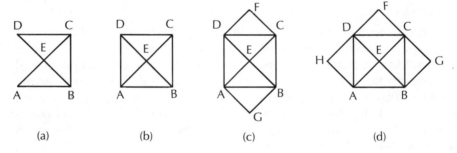

Figure 10–4

Any network with exactly two odd vertices is traversable by a single path— can be drawn without lifting the pen or retracing. One of the odd vertices is the starting point of the path, and the other is the stopping point.

A network which has no odd vertices at all, such as figure c, can be traversed by a single path no matter where one starts.

Now we can return to the Shongo children with the knowledge that the task is not impossible, since each figure has two odd vertices, at A and at B. You may want to start with the following simplified versions of the two Shongo networks (Figure 10–5). Note that the finishing point, vertex B, alternates its position with reference to vertex A, with each successive extension of the network.

Figure	A	B	C	D	E	F	G	H	Number of Odd Vertices	Traceable?
a	2	3	3	2	4	–	–	–	two	yes
b	3	3	3	3	4	–	–	–	four	no
c	4	4	4	4	4	2	2	–	none	yes
d	4	4	5	5	4	2	2	2	two	yes

Table 4. Number of line segments meeting at each vertex.

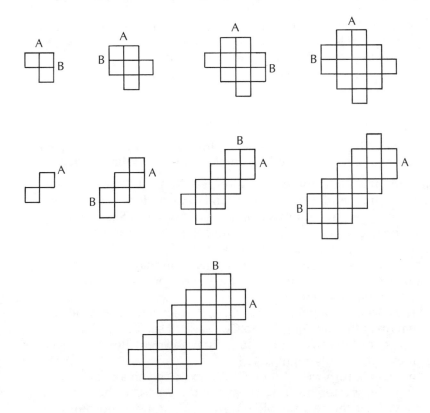

Figure 10–5 Simplified versions of the Shongo networks.

Figure 10–6 Kuba (Shongo) embroidered raffia cloth, Zaire (Congo). The interlacing mbolo pattern is similar to those drawn in the sand by Kuba children. This cloth dates back to the eighteenth century. British Museum.

The very elaborate network of the Jokwe people of Angola (Figure 10–7) illustrates the story of the beginning of the world (Bastin, page 39):

At one time the Sun went to pay his respects to God. He walked and walked until he found the path which led to God. He presented himself to God, who gave him a cock and said to him: "See me in the morning before you leave." In the morning the cock crowed and woke up the Sun, who then went to see God. God said: "I heard the cock crow, the one I gave you for supper. You may keep him, but you must return every morning." This is why the Sun encircles the earth and appears every morning.

The Moon also went to visit God, was given a cock, who also woke him up in the morning . . . So God said: "I see that you also did not eat the cock I gave you for supper. That is all right. But come back to see me every twenty-eight days.". . .

And man in turn went to see God, and was given a cock. But he was very hungry after his long voyage and ate part of the cock for supper and kept the rest for his return trip. The next morning the Sun was already high in the sky when our man awoke. He quickly ate the remains of the cock and hurried to his divine host. God said to him with a smile: "What about the cock I gave you yesterday? I did not hear him crow this morning." The man became fearful. "I was very hungry and ate him." . . . "That is all right," said God, "but listen: you know that Sun and Moon have been here, but neither of them killed the cock I gave them. That is why they themselves will never die. But you killed yours, and so you must die as he did. But at your death you must return here." And so it is. The top figure is God, the bottom is man, on the left is the Sun and on the right is the Moon. The path is the path that leads to God.

Figure 10–7

Riddles

The Kpelle children of Liberia tell the familiar story with their own cast of characters. This exercise in logical reasoning is about a man who has a leopard, a goat, and a pile of cassava leaves to be transported across a river. The boat can carry no more than one at a time, besides the man himself. The goat cannot be left alone with the leopard, and the goat will eat the cassava leaves if he is not guarded. How can he take them across the river?

Since there is no mutual attraction between the leopard and the cassava leaves, they are the only pair that can be left alone together. Therefore the man must first ferry the goat across. Here is one solution. Is there another?

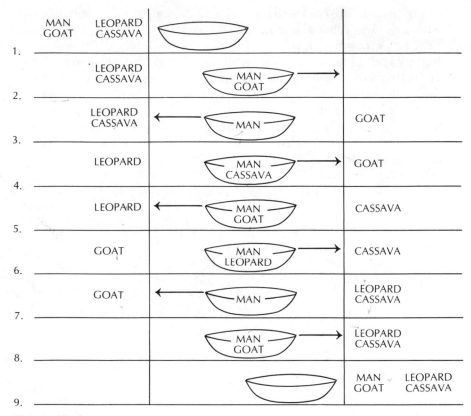

Figure 10–8

Arrangements

Many children's games depend upon an arrangement of objects. The Kpelle line up sixteen stones in two rows of eight each. One person is sent away, and the others choose a stone. When the "out" person returns, he must determine which stone has been selected. He may ask four times in which of the two rows the stone is located. After each reply, he may rearrange the stones within the two rows. He must be able to identify the chosen stone after the fourth reply.

The key to the solution lies in the procedure by which the stones are rearranged each time. After the first reply the questioner rearranges eight stones, half the original number, so that they are now in different rows; after the second reply he interchanges four stones, half the previous number, and the next time he changes the positions of two stones. The answer to the last question determines precisely which stone has been chosen.

Initial Position	Row 1	1	2	3	4	5	6	7	8
	Row 2	9	10	11	12	13	14	15	16

Suppose 6 is the chosen stone. Here is one procedure the questioner may use:

Answer to question 1: "It is in the first row."
He interchanges the odd-numbered stones:

$$1 \longleftrightarrow 9, 3 \longleftrightarrow 11, 5 \longleftrightarrow 13, 7 \longleftrightarrow 15$$

Second Position	Row 1	9	2	11	4	13	6	15	8
	Row 2	1	10	3	12	5	14	7	16

Answer to question 2: "It is in the first row."
He knows that the stone is even-numbered. He interchanges stones numbered $4n$ (multiples of four): $4 \longleftrightarrow 12, 8 \longleftrightarrow 16$

Third Position	Row 1	9	2	11	12	13	6	15	16
	Row 2	1	10	3	4	5	14	7	8

Answer to question 3: "It is in the first row."
The chosen stone must be 2 or 6. He interchanges 6 and 14.

Fourth Position	Row 1	9	2	11	12	13	14	15	16
	Row 2	1	10	3	4	5	6	7	8

Answer to question 4: "It is now in the second row."
The only candidate is stone 6. That, of course, is the correct answer.

Children in northeastern Tanzania play *tarumbeta*, so called because the pattern formed by the beans resembles a trumpet. Forty-five beans are arranged in nine rows to form a triangle (Figure 10–9). Four boys participate. The "chief" sits at the apex, and serves as umpire, while the challenger sits at the base with his back to the triangle. The other two boys sit on each side of the triangle and remove one bean at a time, in turn, starting at the base and working toward the apex. After each move the

challenger must give the number name of the bean just removed. However, whenever the first bean of any row has been picked up, he is forbidden to answer. Of course, he is not permitted to look at the triangle,
but must visualize the position of the beans after each move. Youngsters are trained for such games by their older playmates. Small children
may start with arrangements of ten objects, and work up to the required
number. In the case of ten objects, the challenger calls out: "(silence), 4,
2, 3, (silence), 7, 6, (silence), 8, 10." Now try it with the triangle of forty-
five objects arranged in nine rows!

APEX   APEX

BASE   BASE

Figure 10-9 Children in northeastern Tanzania play tarumbeta, involving a
triangular array of 45 stones. Young children are trained with arrays of ten
stones.

The children must be aware of the triangular numbers, those numbers which represent the sums of the first n counting numbers, an arithmetic series (Figure 10-10).

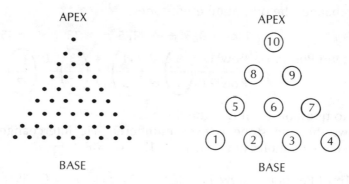

NUMBER OF ROWS: $n = 1$ 2   3   4   5

NUMBER OF STONES:

$$\text{SUM} = \sum_{k=1}^{n} k = 1 \qquad 3 \qquad 6 \qquad 10 \qquad 15$$

Figure 10-10

Σ is the Greek letter sigma, representing "sum."

$\sum\limits_{k=1}^{4} k$ means "the sum of the consecutive integers from one to four," or $1 + 2 + 3 + 4 = 10$.

Games of Chance

Adult men in most of Africa play gambling games based on the outcomes when chips, nuts, or cowrie shells are tossed. These are similar to our games of dice and coin-tossing. The procedures frequently resemble those of the diviner. Indeed, the position of the gambling objects establishes the gambler's relationship to the controlling powers in the society, and the play may be preceded by certain rituals to invoke a favorable outcome.

In one version of the Igbo game *Igba-ita*, freely translated "pitch and toss," four cowries are thrown; the most favorable outcomes are all four up or all four down, just as in divination (see Chapter 5). More recently coins have replaced cowries, and the name has been changed to *Igba-ego* (*ego* = money).

What are the chances for an outcome in which all four land in the same position? Prediction is easy in the case of coins. There are sixteen different possible outcomes when four coins are tossed, as follows (H = head, T = tail):

HHHH	THHH	THHT	THTT
HHHT	HHTT	THTH	TTHT
HHTH	HTHT	TTHH	TTTH
HTHH	HTTH	HTTT	TTTT

Because of the symmetry of a coin, a head or a tail is equally likely to occur, and each of the sixteen outcomes is as probable as any of the others. In the long run one would expect that two out of sixteen, or one out of eight tosses would result in an outcome in which all four coins land in the same position. A similar prediction cannot be made for the asymmetrical cowrie; further research would be required before giving odds on four of a kind. No doubt the Igbo men had it down to a science!

In earlier times a favorite market-day recreation was this game of *Igba-ita*. From two to a dozen might participate in this version described by Basden (1966, pages 352–353). The players squat in a circle, each with a heap of cowries in front of him to serve as a bank. The challenger picks up twelve cowries, and the players stake six, twelve, or more shells. The challenger tosses his handful so that the shells spread as they fall. Winning or losing depends upon the combination of cowries falling with the openings up or down, somewhat analogous to heads and tails on coins. The winning combinations are all twelve alike, six up and six down, or eleven one way and one shell the other way. A combination of four and eight signifies that the challenger must yield to another man.

"Quick as lightning," writes Basden," the players note the positions and forfeit their stakes or collect their gains. The play becomes exceedingly fast, and soon a cloud of dust encircles each group of gamblers. I have watched players at this game, and it has always been quite beyond me to note the positions of the fall; the cowries have been counted and snatched up again long before I could begin to count."

In the Yaounde society of Cameroon, gambling is an important method of gaining material wealth and improving one's status, according to Morris Siegel. In "A Study of West African Carved Gambling Chips" he describes the Yaounde game called *Abia*, played both with discs of calabash bark and with gambling chips. The latter, made from the shells of certain nuts, are beautifully decorated on the outer side, called the right side (Figure 10–11).

Figure 10–11 Beti (Yaounde) carved gambling chips, representing a lizard and a snake. Several dozens of these chips are in the Schomburg Collection of the New York Public Library. (See Frederick Quinn: "Abbia Stones," African Arts (UCLA) IV, No. 4 (Summer 1971), pp. 30–32.)

This is Siegel's description of one version of *Abia*. The players sit in a circle, each with his own carved chips and a supply of small iron bars or other currency. Each player contributes a chip, which is placed on a woven plate together with seven discs. An impartial person, the arbitrator, raises the plate and then flings it to the ground so that its contents are now covered by it. After the players have placed their bets, the plate is removed, and the positions of the discs and chips are analyzed as to the number of each with the outside (right side) uppermost. The winners are determined by a set of complicated formulas involving the ratio of chips and discs falling in the right or wrong positions.

For example, suppose six men participate. Here are several possible outcomes:

1. If two or four chips fall right side up, along with two or three of the seven discs, these chips are the winners.

2. If one chip alone falls in the same position as all seven discs, that chip wins.

3. If two chips land in the same position as all seven discs, these two are losers.

4. If all the chips land with the outer side up, but all the discs have their backs down, the play is invalid.

Turnbull tells how the Mbuti (Pygmies) of the Ituri forest, in the Congo, used a game of chance to outwit the villagers who used their camp as a convenient stopping place. When the villagers asked for food, the Mbuti challenged them to a game of *panda*. A specified number of beans are thrown on the ground, and one of the players scoops up a handful. His opponent estimates at a glance how many are needed to make up a multiple of four, and asks the first man for one, two, three, or zero beans. "The count is then made—also at a glance—and if the player has an exact multiple of four he wins" (page 138).

Winning in *panda* depends less upon chance than on a knowledge of addition modulo four. The player mentally groups the beans in fours, and then considers only the number left over, the remainder.

Turnbull claims that the Mbuti always win against the villagers. "For one thing they have an uncanny sense of form that enables them in a moment to know the exact number thrown down." Moreover, when they see that the opponent is about to win, they are adept at dropping an additional bean concealed between the fingers or in the hair.

CHAPTER 11
THE GAME PLAYED BY
KINGS AND COWHERDS—
AND PRESIDENTS, TOO!

Several decades ago there lived in the western part of Tanganyika (now Tanzania) a youngster named Kamberage. He delighted in playing *soro* with his father's friend, Chief Ihunyo, and no wonder, since the boy always came out the winner in this complicated African board game. Much impressed by Kamberage's skill, the older man advised his friend, Chief Nyerere, to send his son to school. Nyerere accepted the suggestion, and soon Kamberage was enrolled in a Native Authority boarding school in Musoma. In no time at all he was at the head of his class.

No doubt Julius Nyerere, the president of Tanzania, would have attended school even if he had not excelled in *soro*. As a college student in Scotland he became a champion chess player. Even today the Tanzanian president's house is the scene of frequent *soro* tournaments, and *Mwalimu*, the "Teacher," as the Tanzanian people like to call him, is hard to beat.

In past centuries each new Kabaka of Buganda, upon his accession, was required to perform the ritual of *okweso*. The new king picked brown seeds from a certain tree, and these were later used in the *omweso* board game kept in the royal hall. Here the Prime Minister played *omweso* while he decided legal cases. The significance of this ritual is that the Kabaka shall not be outwitted by his people. If they should try to trick him, he would overcome them by his strategy, just as an *omweso* expert defeats his opponent by a clever move.

In the British Museum is a beautifully carved wooden statue of Shamba Bolongongo, ninety-third king of the Kuba people of the Congo (Figure 11–1). In front of the seated king is a game board, with parallel rows of circular troughs, and a larger semicircular bowl at each end. It is said that this statue was carved from life in about the year 1600, the beginning of a tradition of representing each Kuba king with his most noteworthy accomplishment. King Shamba's people were called the Shongo because of their deadly throwing knife. He induced them to give up their war-like practices and to devote themselves to building their country. Shamba introduced the cultivation of new crops, the arts of wood carving and weaving, and he also extended trade with other peoples. Above all, he taught his subjects the board game called *lela*, persuading them to substitute this intellectual pastime for the gambling games at which they had whiled away much of their time.

AFRICA COUNTS

Number and Pattern in African Culture

Figure 11–1   King Shamba Bolongongo. British Museum.

In many parts of the continent, game "boards" have been cut into rock. Some of these are of great antiquity, others are recently made. A middle-aged man of Uganda related that in his youth, he and his friends would pass the time playing *omweso* while they tended the cattle and goats. With a hard pointed stone he could hollow out the four rows of depressions in a rock surface in a short time. The boys would play with any available beans or pebbles. Often youngsters would improvise game boards by scooping out holes in the earth.

One of my friends spent a summer in Surinam, South America, living with the descendants of African slaves. In the eighteenth century their ancestors had fought for and won their freedom from their Dutch and English captors. Jan could not speak their language at the beginning, but she could learn to play *adji boto*. For the first few days of her visit her hosts communicated with her through the board game, played with *adji* beans on a board having two parallel rows of six holes.

The world's oldest game has hundreds of names, dozens of versions of play, and is popular throughout Africa, as well as in parts of Asia, in the Philippines, the West Indies and South America. It is played by kings on beautifully carved ivory boards embellished with gold, and by children in holes scooped out of the earth, using pebbles or seeds as counters. Now even computers are playing this game, frequently referred to by its Arabic name *mancala*, or "transferring." By programming electronic digital computers to play games, engineers at the Massachusetts Institute of Technology are learning how machines can make decisions. There are fewer possible arrangements of the pieces in the simple versions of mancala than in chess or checkers. If mancala is played with just thirty-six counters distributed in two rows of six holes, there are only about 10^{24} (a million times a million times a million times a million) total possibilities. The computer is instructed to analyze the relative advantages of all possible moves. This may take several minutes, even for a fast computer. But it can be programmed to consider only the advantageous moves, so that instead of analyzing 414,000 positions before deciding on a move, it need now consider less than 4000.

Two-Row Versions

North of the Equator the two-row board is used almost universally. In the south and east, the four-row game board is generally used, and in Ethiopia some people play on a board of three rows. Here I shall describe some of the two-row versions of the game.

R. C. Bell, the British authority on the history of games, has ranked a version called *wari*, or *oware*, among the world's nine best games. The Asante people of Ghana play it on a board with two rows of six holes, and an additional hole, or "pot," at each end (Figure 10–1). The two players face each other and begin by placing four seeds in each of their six holes. A player's home territory consists of the six holes in the row on his side and the pot at his right.

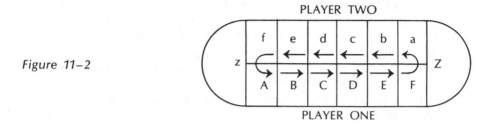

Figure 11–2

In his first move, Player One scoops up all the seeds from any hole on his side and "sows" one seed in each hole in a counter-clockwise direction, beginning with the hole adjoining the empty one. Player Two does the same with four seeds from one of his holes, and so they continue.

The object of the game is to capture at least twenty-five seeds, a majority of the total forty-eight, in one's own pot. If a player, when distributing the contents of one of his cups, has dropped the last seed in an enemy cup which already contains one or two counters, he may capture the two or three seeds in this cup, as well as the seeds in all enemy cups having two or three seeds, going clockwise in an unbroken sequence. At the end of this chapter some of the plays will be illustrated in greater detail.

If a player finds that the opponent's cups are empty, he must try to feed his seeds into enemy territory, if possible. If this is not possible, he gains all the remaining counters. A few final pebbles circulating endlessly around the board are divided between the two players, each going to the player who is moving it.

The Yoruba version of the game is called *Ayo*, and is played on a board, *Opon Ayo*, having two rows of six holes. The counters are the seeds of a certain plant that grows wild in Nigeria, greyish in color and almost spherical. Four men, *Omo Ayo*, are placed in each hole. The rules are similar to those of the Asante game *wari*. When the opponent's holes have been cleared or have only one "man" in each hole, the game has ended.

Adi is the name of the game in southern Ghana, where it is played with *Adi* seeds gathered from the *Aditi* bushes. This version, like those already described, begins with four seeds placed in each of the twelve "houses," arranged in two rows. At each end of the board is a "treasury." Player One takes all four seeds from any house on his side and drops them, one by one, in each hole in a counter-clockwise direction. If the last seed falls in a house already containing seeds, he picks them all up and continues around, each time picking up all the seeds from the hole in which he has just dropped his last seed. His turn ends when the last seed in his hand falls into an empty hole. Then it is the turn of Player Two.

If a player makes a house of four as he drops his last seed, he takes that group, whether it is on his side or his opponent's. He may also capture any house of four that appears on his side of the board, even if it occurs during his opponent's turn. He places the captured seeds in his "treasury." The wary player tries to prevent groups of four from appearing on his opponent's side.

The winner of the round is the player with more than enough seeds in his treasury to fill his houses with four in each. He uses his surplus to buy empty houses from his opponent by filling them with his own seeds. The players then start the next round. They proceed as before, except that the groups of four that appear in the purchased houses now belong to the buyer. The object of the game is to buy up all of the opponent's houses. My experience with this version is that the rich get richer. Once I had lost any of my houses to my opponent, I was never able to win.

Figure 11–3   Board for the Yoruba game Ayo. Dahomey. British Museum.

It is amazing to discover how many varieties of the game can be played on a board which has two rows of six cups each and a large hole at each end for the captured pieces. The anthropologist, Stewart Culin, set down the version described to him by Prince Momulu Massaquoi, son of the king of the Vai people of Sierra Leone and Liberia. They call the game *Kpo*, in imitation of the sound made by the seeds or ivory balls as they are dropped into the cups. Two, three, or four players may compete, dividing the cups equally among them. When two play, four beans are placed in each hole. However, when three or four play, they start with three beans in each hole, and they may distribute the counters in either direction.

An uneven distribution of beans in the opening move is a feature of the Igbo versions, called *azigo* or *okwe*. It is played with black spherical seeds of the *okwe* tree, on a board of two rows, having from ten to twenty holes per side, in one version described by Basden. Two, three, or four players may participate. A player lifts the beans from a cup and distributes them one at a time in the holes; his aim is to have the last one drop opposite an enemy hole containing exactly one or three beans, which he may then appropriate and use in play. Basden says: "A clever player will calculate numbers and spaces at an extremely rapid rate; a weak opponent is apt to lose his original counters in an incredibly short time, which is either very humbling or exasperating, according to his temperament. A white man usually has no chance against a good native player" (Basden, 1921, page 136). Occasionally the game led to excessive gambling and to quarrels. A simpler version of the game is played on a board having six holes on a side.

Commented Mbonu Ojike, the Igbo writer: "It is an adult's game, and when it is in process no one talks. Both the competitors and spectators remain mute, working with their eyes, heads, and hands. Sometimes a champion is tied up until his wife comes, overturns the board, and says: 'Leave that nonsense thing and come home for dinner'" (page 141). In his community the forty-hole board was used, and two played at a time. The distribution of seeds was almost identical to that described by Basden for the opening of the game. When a player's last seed landed in a hole in his own territory, opposite a cup containing three seeds, he was entitled to "eat" the enemy's three seeds.

The Kikuyu call their game *giuthi*, again played on a board with two parallel rows of six holes and a "shed" for captured seeds at each end. Normally two people play, but it may also be played by teams of three or more. The game starts with an equal number of seeds in each hole. The number may be from four to nine, and the first player has a choice of going to the right or the left as he distributes the contents of any cup, one by one, into the next consecutive cups. Then he picks up the entire contents of the cup into which he has placed the last seed, and drops them in the opposite direction. He continues in this fashion, reversing the direction each time, until the last seed in his hand falls into an empty cup. If this last play ends in the opponent's territory, it is the opponent's turn to play. If the last play ends on the player's own side, after he has traversed enemy territory, he may capture the contents of the opponent's hole opposite his empty one or all of the enemy cups opposite a sequence of his empty holes. The winner is the one who has captured the largest number of seeds.

The directions just given are for the simpler version of *giuthi*, or part one of a complete game. The more complicated second part is played between the "rich man," the winner in part one and the "poor man," the loser, and requires that a player capture all of the opponent's seeds, by setting "tricks."

Perhaps there are no African folk who do not enjoy some version of the game. The Kenya Luo call it *ajwa*, and play on a board having two rows of eight holes. Three seeds are placed in each hole, and the game opens with each player picking up the seeds from one of his holes and dropping them, one by one, in successive holes, going around the board.

The Kamba of Kenya play on a board of two rows with ten to twenty holes in a row and start with four pebbles in each hole. Their rules are similar to those of the Kikuyu. The Maasai game may have up to fifty holes in a row.

Playing on a Four-Row Board

In the southern and eastern parts of the continent, the four-row game board is popular. *Bao*, meaning "board," is the Swahili name, but there are dozens of other names and numerous methods of play. The version played in parts of Uganda, called *omweso* or *mweso*, requires sixty-four beans on a board having four rows of eight holes. Each of the two players makes a circuit of his own two rows only, as though each contestant were playing on his own two-row wari board. The game opens with each player distributing his thirty-two seeds as he chooses in the sixteen holes on his side of the board. Each player in turn sows his beans, one by one, in a counterclockwise direction in his two rows after some opening moves. If the last bean falls into an occupied hole, he picks up the contents and continues his sowing. If the last bean falls into an occupied hole in his inner row, the player may capture the contents of the two holes in enemy territory opposite to his hole, provided they both contain counters. Under certain conditions a player may make a reverse move — in a clockwise direction — in order to capture his opponent's beans.

In *omweso* all sixty-four beans remain in play throughout the game. The captured beans are not set aside, but are sown one by one in consecutive holes. Part of the excitement lies in watching as a player with very few beans left makes those dramatic moves that enable him to win the game. At the end of this chapter are complete rules for playing *omweso*.

Bao kiswahili, the east coast version, is similar to *omweso* in that the board has four rows of eight holes, each player has thirty-two beans, and the object is to capture the opponent's men and redistribute them on one's own half of the board. As in *omweso*, if the last bean falls in an unoccupied hole, the move ends, but if it falls in an occupied hole, the contents are picked up and the move continues. However, in the opening moves of *bao kiswahili*, each player uses just ten beans, reserving the other twenty-two for the middle play. Captures can be made only from the front row, and captured beans are always sown from the end of the row, not from the adjacent hole. One aspect of the strategy is to build up an accumulation in the fourth inner hole, which is usually of a different shape. A player loses the game when the holes in his inner row are empty or contain at most one bean in each.

Figure 11–4 Board for Omweso (or Mweso), played by the Ganda of Uganda. British Museum.

Further south and west the game is played on boards of four rows, but with a variety of from six to twenty-eight holes in a row. When Torday visited the Kuba (Shongo) of the Congo area early in the twentieth century,

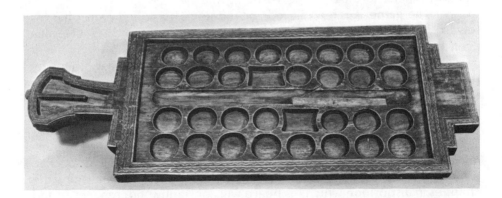

Figure 11–5 Board for Bao, Swahili, East Africa. Note the two square holes. British Museum.

he found that they played *lela* on a board with four rows of seven holes, and an additional larger hole at each end. Each player placed nine seeds in the first hole, starting at the left of the row closest to him, six in the second hole, two in the third, and one in the fourth. Three seeds remained in the player's hand, and could be dropped, one by one, in any three consecutive holes. Torday remarked that they played very quickly, and frequently a player hid his part of the board to prevent the moves from being seen. The game, he added, was rarely played, and one found skilled players only among the aged. King Shamba Bolongongo would have been saddened.

Customs and Mores

Although the Kabaka of Buganda is said to have played the game with his wives and sisters, it is rarely played by men and women together. Apparently no man's reputation could bear the ridicule that would be heaped upon him if he were defeated by a woman! Ganda women outside of the royal court were not allowed to play for fear their crops would fail. Young girls were warned that their breasts would not develop, and no men would marry them. Thus the men were assured that the game would not distract the women and girls from their assigned chores in the fields and the home.

I asked several people whether women play in their communities. Igbo women, I was told, rarely play, but not because it is forbidden to them, since boys and girls do play the game together. However, an Igbo man never plays with a woman. Secondary school students from western Tanzania remarked that it was played by old men who had nothing to do. "Women? Oh, they are not interested in it."

One of the few references to female participation is in the game *ikiokoto*, popular among Ijo women and girls of the Niger Delta. The rules are similar to those of the Yoruba game *ayo*.

The game is rarely played for money. The skilled player has his reward in bringing renown to his family and to his entire village. He may even be celebrated in song!

Every observer has remarked upon the speed and skill exhibited by the contestants. One is not allowed to pick up the contents of a hole to count. The good player remembers the number of seeds, not only in his own holes, but in those of his opponent. Says Driberg: "A move may, as we shall see, consist of several circuits of the players' board and at each circuit the number of marbles in their respective holes varies. In considering the various moves possible to him and in assessing their respective merits, a player has to look several circuits ahead and has to bear in mind the changes in disposition which his board will constantly undergo. Yet a good player can do this almost instantaneously, and there is little pause for consideration between moves" (page 168). Matthews carries his praise even further: "A practised player will not only be able to tell at a glance exactly how far his stones will take him, but can also forecast the moves which he will thus force his opponent to make."

Commenting about *igisoro*, the version popular in Rwanda, Alison Des Forges writes: "It is still regarded as a supreme test of a man's wit and cunning. The game is played at lightning speed and usually draws rapt attention from the onlookers." It is not often played openly in public today, since it is associated with the pre-revolutionary kingdom.

Chief Ayorinde informed me about the customs and etiquette of the game among the Yoruba. It may be played indoors or out, usually after work, but rarely at night. Some chiefs and elders are willing to spend any amount of money to invite players of high repute to their homes for a few matches. He remarks about two methods of play; it may be played just for fun and relaxation, involving little drain on the intellect, or it may be a game of great skill, "as interesting to watch as watching a man who knows his trigonometry inside out and is proving a problem with ease." He continues: "Many players have a code for counting the beans in their holes. This is to baffle their opponent in the game." The code words are related to the rhythm of the Yoruba words. For example, the translation of one set of code words is:

2	A lying
2	woman
2	Proposals were made to her
2	She refused
2	When presented with money
1	She came home.

Part of the pleasure comes from playing on a beautifully carved board. Some families possess boards dating back several generations. Even though most people do not own the golden boards of the Asante kings, or the ivory and gold games of the Vai chiefs, they take pride in their very artistic boards of wood. At the opposite extreme, youngsters out minding the goats, or young men with a few minutes off from work, dig holes in the earth, gather the required number of pebbles or seeds, and have a game going in short order.

What is the connotation of the terminology used for the counters and the pits in different cultures? Many versions refer to war— "captives," "kill," "you cut his throat." The Kikuyu counters represent "cattle" in their "fields," and the captured pieces are placed in "sheds." Their game is usually played out of doors by old and young Kikuyu men and by girls while looking after their herds of cattle, goats and sheep. To these herders, cattle represent the most prized possessions. In the game of *adi* the participants "buy houses" and place their wealth in a "treasury." The *wari* player "sows seeds" and stores the captured counters in a "pot."

History of the Game

The game is of great antiquity. There is some evidence that it was invented by the Sumerians several thousand years ago as a system of record keeping having debit and credit entries. One side indicated money or goods received, and the other recorded sales or payments. A limestone "board" surviving from ancient Egypt had three rows of fourteen holes and a storage pot. Several sets of deeply cut holes have been discovered in the roofing slabs of the Kurna temple at Thebes, in the summit of the great pylon at the entrance to the temple of Karnak, and at the Luxor temple. Rock cut boards of great antiquity have been found at Zimbabwe, in Uganda, in Ghana, and in many other areas of Africa.

In parts of Ethiopia there are whole fields of giant stone megaliths, from ten to fourteen feet high, in the shape of phalluses. No one knows how long they have been there or who erected them. Many have fallen over, and are used as benches. In one of them a game board has been cut!

Alvarez, the Portuguese ambassador to Abyssinia (Ethiopia) from

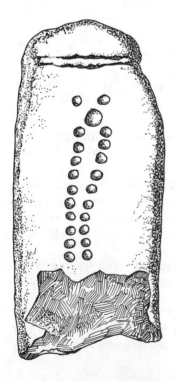

Figure 11–6   A game "board" has been cut into this giant stone megalith in the shape of a phallus, one of many of unknown origin found in Ethiopia. (After Jensen.)

the years 1520 to 1527, referred to the game called "mancal," already anti-quated at that time. A more recent Ethiopian version, *gabatta*, was described by Bent in 1873. The board, made of wood or cow dung, depending upon the resources of the owner, has eighteen holes, nine for each person. "There are three balls, called chachtma, for each hole, and the game is played by a series of passing, which seemed to us very intricate, and which we could not learn; the holes they call their toukouls, or huts, and they get very excited over it" (quoted in Culin, page 601, footnote).

Figure 11–7   Board for gabatta, Ethiopia. British Museum.

A European visitor to Gambia in the seventeenth century was much impressed by the skill of the contestants. "Some of them are wondrous nimble," he said of the men who played under shady trees in the heat of the day.

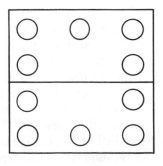

Figure 11–8   Centuries ago the members of the royal court of Benin, Nigeria, played on bronze game boards constructed on patterned pedestals. (After Luschan.)

Such was the wealth of the Asante kings that they played *wari* on gold "boards," fashioned in the shape of the traditional sacred golden stool, symbol of the authority of the Asantahene, as the ruler was called. Even today the wari board is made with a pedestal in which the seeds may be stored when not in use. Some of the beautiful bronze weights of the Asante were formed in the shape of wari boards (Figure 6–1).

Nsimbi tells the history of *omweso* in Uganda. Originally it was the principal pastime in the courts of the king and the various chiefs throughout the country. "The game provided chiefs with the opportunity to meet their people socially and to know them personally. As they played the game they could talk about different topics, thereby getting to know what was going on in the country" (page 9). A certain prime minister was reputed to have the ability to "play the game and at the same time listen to the statements of two people in a suit. At the end of the statement he could repeat the plaintiff's and the defendent's statements with amazing accuracy and then give his judgement, which is said to have been often correct and fair" (page 10).

By the end of the nineteenth century the popularity of the game was waning, and with good reason. After Uganda had become a British protectorate in 1894, all able-bodied men were required to devote some time to carrying the heavy loads of the government officers. One way to avoid such work was to stay away from the chiefs' palaces. When cotton-growing was introduced to enable the peasants to pay their taxes to the British, men were forced to engage in agricultural labor full time. The children began to attend school, and the game was not encouraged by the short-sighted educational authorities. Another drawback lay in the fact that many men spent so much time traveling long distances from their villages to the towns in which they worked, that they had neither leisure time nor energy for the game.

The revival today is due in part to improved conditions of living and working, and in part to the great interest Africans have in reviving their own institutions, now that the colonial period is coming to an end.

Recently the National Museum in Dar es Salaam, Tanzania, featured an exhibit of *bao* boards. One had four parallel rows with eight holes in each. Another was an ancient board made by the Makonde people, refugees from the Portuguese colony of Mozambique, and had two rows of eight holes. A contemporary Makonde artist had carved two tiny wood figures, just a few inches high, crouched on either side of a *bao* game. A clipping from a Dar es Salaam newspaper, the *Sunday News*, contained a photograph (Figure 11–9) with the caption: "Bao experts Maalim Jusuf Alawi of Mtwara and Mzee Bin Salim of Dar es Salaam are pictured during a preliminary session in preparation for Monday when they will be teaching the game to members of the public for a week at the headquarters of the National Development Corporation. The bao lessons have been organized under the auspices of National Arts of Tanzania, an N.D.C. subsidiary which is now holding an exhibition of Makonde wood sculpture at Development House on Kivukoni Front, Dar es Salaam" (August 2, 1970). Several youngsters were having an animated conversation in Swahili as they stood in front of the display case. I asked them whether they could play. They were obviously amused, and replied: "Everyone plays bao!"

Figure 11–9 Game of bao. This photograph appeared in the Sunday News, *Dar es Salaam, August 2, 1970, with the caption: "Bao experts . . . are pictured during a preliminary session in preparation for Monday when they will be teaching the game to members of the public for a week . . . under the auspices of National Arts of Tanzania . . ." Courtesy of National Development Corporation, Tanzania.*

In the New World, too, the game is undergoing a revival. It was known many years ago in Bahia, Brazil among the descendants of the Yoruba people, and then forgotten until recently when it was rediscovered by students at the Centre of Afro-Oriental Studies in the University of Bahia.

One method used by social scientists to trace the history of African peoples is to investigate their style of playing this game. Some people learned it from their neighbors of the same ethnic background, while others picked it up from their conquerors or their subjects. This was of particular value in tracing the descent of black people now living in the New World. In Surinam the game is associated with death; it is played in the House of

Mourning to amuse the spirit whose body lies awaiting burial. The board must be carved by a man who has attained old age and has lost a wife.

I used the game to do some detective work on my own. The setting of Nadine Gordimer's fascinating novel *A Guest of Honour* is an unspecified African country. As I was reading the book, it seemed to me that any one of several countries could have been the locale. My clue was the description of a game of *chisolo*, played in holes scooped out of the dirt. Immediately I recognized the country as Zambia.

Do African lions play the game? Some people think so! In her book *Born Free*, Joy Adamson describes her experiences in tracking down two man-eating lions. These beasts had killed or mauled about twenty-eight people of the Boran tribe in northern Kenya. The Boran are very brave and are not afraid to hunt lions with a spear. But in this case they were overwhelmed, in part because of the tactics of these two lions, and in part due to a superstition. "It was said that before starting off on a raid the lions would repair to an open sandy place and there make two rows of depressions in the sand with their paws. Then, using twigs as counters, they would play the ancient game of *bau* (a game of unknown antiquity, which resembles draughts and is played all over Africa). If the omens were good they would raid a *boma* and claim a victim; if not, they would wait" (page 68).

Educational Value

The pit and pebbles game is probably the most arithmetical game with a mass following anywhere in the world. In its simplest two-row form it can be played purely as a game of chance by very young children. Even in this version it has educational value in encouraging the child to count. He learns the concept of a one-to-one correspondence as he drops each of his counters into each of a sequence of consecutive holes.

Soon he learns simple sums. For example, if Player One has nine beans in pit B, he sees that he will drop four in his territory and five in the opponent's, thus bringing him to hole e. Then he may take short cuts in planning his strategy. If he is removing beans from hole B, he recognizes that he needs to sow six beans to bring him to hole b, the opponent's hole corresponding to his own starting position. When he picks up nine beans from hole B, he automatically subtracts six from nine and knows he will advance three holes from hole b to reach hole e. A move of eleven means he will drop his last bean in the hole just preceding his starting position, the kind of addition useful in reading the clock.

Omweso, which permits reverse moves, introduces the concept of negative numbers. As he works with the sixteen holes in his own territory, the player sees that a backward move of six spaces is the equivalent of a forward move of ten and that a move of (-3) brings him to the same position as a move of $(+13)$.

It is incredible that these African games were actually discouraged by the colonial education authorities in favor of ludo, snakes-and-ladders, and similar games of European origin. I agree wholeheartedly with John B. Haggerty, who writes: "Kalah [an American version, commercially produced] is the best all round teaching aid in the country" (page 328), and later: "In addition to its value as a diversion and as a means of developing the intuitive abilities so important to problem-solving, there is another outcome equally valuable. This outcome is the recognition of the close identification of the game throughout the history of civilization with the development of systems of numeration and the concept and ideas of number" (page 330).

The game is being promoted in the United States, where several versions have been marketed commercially. In 1891 the Milton Bradley Company issued Chuba, played with sixty counters on a four-by-eleven-hole board, but it soon died a quiet death. All of the current versions are played on boards having two rows of six holes. In Pitfall, issued by Creative Playthings, the player who has dropped his last seed into an empty cup on his own side then captures the seeds in his opponent's cup directly opposite. It is similar to Kalah, designed in 1940 by William Champion and now manufactured by Products of the Behavioral Sciences. Oh-Wah-Ree, marketed by the 3M Company, can be played by two, three, or four players. The instruction booklet gives the rules of play for six varieties of the game. Pendulum Oh-War-Ree, for example, calls for a change in direction for each lap of a player's turn, as in the Kikuyu version.

Figure 11–10 Pitfall is one American version of the African game. Courtesy of Creative Playthings.

A Sample Game of Wari

(Based on the sample game of Standard Oh-Wah-Ree described in the Oh-Wah-Ree instruction booklet)

Each diagram shows the position of the counters after the execution of the first move below the diagram. A rectangle indicates the counters captured in the first move.

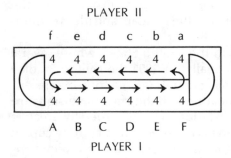

1. Initial position.

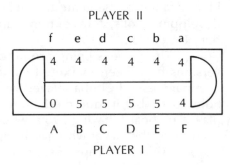

2. Player I sows seeds from cup A.

3. Player II empties cup f.
   Player I sows 6 from cup B.
   Player II sows 4 from cup e.
   Player I sows 8 from cup C.

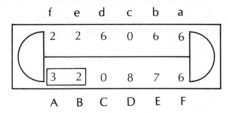

4. Player II sows 5 from cup c and captures seeds in cups B and A.

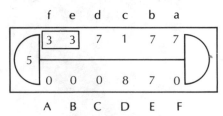

5. Player I sows 6 from cup F and captures seeds in cups f and e.
   Player II sows 7 from cup a.

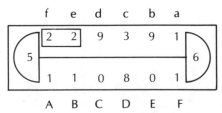

6. Player I sows 7 from cup E and captures seeds in cups f and e.
   Player II sows 9 from cup b.

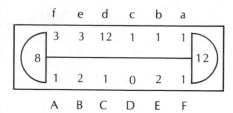

7. Player I sows 1 from cup F and captures seeds in cup a.
Player II sows 4 from cup c and captures seeds in cup A.

8. Player I sows 9 from cup D (poor move!)
Player II sows 3 from cup f and captures seeds in cups C, B and A.

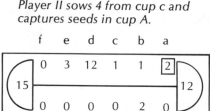

9. Player I sows 1 from cup F and captures seeds in cup a.
Player II sows 1 from cup c.
Player I sows 2 from cup E.
Player II sows 1 from cup a.
Player I sows 1 from cup A.
Player II sows 1 from cup a.

10. In the American version, this ends the game. Player II is the winner, 34 to 14.

Figure 11–11

Rules for Playing Omweso (Mweso)

(based on *Omweso, A Game People Play in Uganda*, by M. B. Nsimbi)

Board: four rows, eight holes in each row.

Players: two (or two teams), one on each side of the board, each player controlling the sixteen holes in front of him.

Counters: sixty-four seeds or pebbles, thirty-two for each side.

Opening Stage:

1.   Each player arranges the counters as in Diagram 1, four in each hole of the outer row, to check that he has thirty-two.

2.   Any side may open. After the first game the loser normally starts the next game.

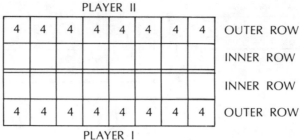

Diagram 1.   Preliminary position, to check that each side has 32 counters.

Diagram 2.   Nninya—The Fourteen. One of the conventional groupings for the opening play.

(Steps 3, 4 and 5 may be omitted by novices.)

3.   Each player places his 32 counters in one of several conventional groupings (described in Nsimbi's booklet) in his 16 holes, so that they are in the most strategic positions (see Diagram 2). The aim of this grouping is to arrange the counters into relay positions from which the players can make a move of more than a full round of the 16 holes.

4.   In turn, each player drops his counters into his holes according to the rules for the particular conventional grouping that he has chosen.

5.   As soon as one player has captured some counters from his opponent, counters are dropped according to the normal rule, one counter in each hole.

Moves:

1.   Each of the two players in turn scoops up all the counters from any hole on his side containing two or more counters, and distributes them in a counter-clockwise direction (see Diagram 3), one in each hole, starting with the hole adjacent to the one from which he had picked up the coun-

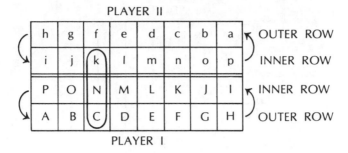

Diagram 3.   Capture by Player II.

ters. If the last counter falls into an occupied hole, the player picks up all the counters in that hole and continues dropping them as before. He has completed his move when his last counter falls into an empty hole. (Nsimbi compares a move to a relay race.) Then it is his opponent's turn to move.

2.   Each player tries to capture as many of his opponent's counters as possible, while protecting his own. All 64 counters remain in play throughout the game.

Capturing Counters:

1.   Vulnerable Position. When a player has counters in two opposite holes (e.g., holes C and N in Diagram 3), the counters in both holes are open to capture by his opponent.

2.   Method of Capture. When a player, in the course of his move, drops his last counter into a non-empty hole in his inner row, directly opposite the vulnerable counters in his opponent's holes, he may collect the enemy counters, and then continue his move in the usual manner. He distributes the captured counters as though they were his own, starting with the hole next to the one from which he had made the capture. In Diagram 3, Player II has dropped his last counter in hole k, which already contained at least one counter. He captures the contents of holes C and N, drops them one by one, starting in hole l. A player may make several captures in one move.

3.   Capture by Reverse Move. A player may capture exposed counters in a reverse move, by dropping his own counters in a clockwise direction from any of the holes labelled A, B, O or P (a, b, o or p) in Diagram 3. The maximum number of counters distributed in any single reverse play may not exceed nine. The captured counters may then be used in another reverse play to effect another capture, or in a forward play, starting from the hole next to that from which the reverse move was made.

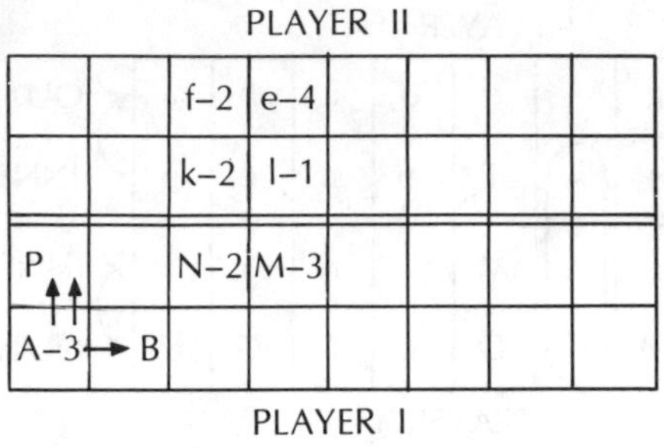

PLAYER II

		f–2	e–4				
		k–2	l–1				
P		N–2	M–3				
A–3	B						

PLAYER I

Diagram 4.   Reverse moves by Player I.

Example of a Reverse Move. In Diagram 4, Player I distributes the three counters in hole A in a reverse (clockwise) direction, starting in P and ending in N. He captures the four counters in k and f, and in another reverse play he drops them, starting in P and ending in M, to capture the five counters in holes l and e. He moves in a forward (counterclockwise) direction with these five, starting in hole B.

4.   Penalty for Failure to Capture. If a player fails to take advantage of an opportunity to capture his opponent's counters during a forward move, the opponent may either group all these counters in one of the two holes, or he may force the first player to take them by making the remark, "Do not leave dead bodies behind."

Victory:

A player wins when he has captured all the counters, or when his opponent has not more than one counter in each hole, thus preventing any move on his part.

These are the basic rules for the simplest version of *Omweso.* The novice may omit the opening stage, and distribute his beans initially as he pleases, thereafter dropping them in the usual manner.

Nsimbi, in his well-illustrated booklet, explains the conventional opening groupings, and several additional methods of winning. Even his version is a simplification of the traditional game. He notes (page 13): "In the past one game used to last between ten and twenty minutes, but now it lasts between three and seven minutes."

The construction of magic squares is a mathematical recreation based on number theory. Their magic lies in the mystifying arrangement of the numbers so that the sum of each row, each column, and each of the two diagonals is constant. Figure 12-I is a third order magic square (three rows and three columns) having fifteen as the magic constant.

4	9	2
3	5	7
8	1	6

Figure 12–1   Third order (three rows and three columns) magic square. The sum of the numbers in each row, column and main diagonal is fifteen. This sum is called the magic constant.

A Chinese myth claims that the magic square called Lo Shu appeared on the back of the divine turtle in the River Lo about 2200 B.C. The Lo Shu (Figure 12–2) is a square array of numbers, later represented by knots in strings, black for even numbers, considered female by the Chinese, and white for the odd (male) numbers. It had a prominent place in ancient Chinese divination.

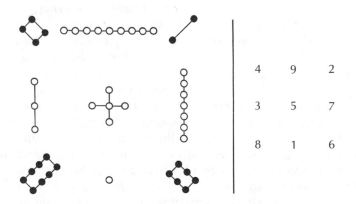

Figure 12–2   Lo Shu, the oldest known magic square, is said to have appeared on the back of a divine turtle in a Chinese river, over 4000 years ago.

Outside of China the earliest writer to discuss magic squares was the Arabian mathematician, Tabit ibn Korra, in the ninth century. About five hundred years later they were introduced in Europe, where they engaged the attention of such illustrious men of science as Cardan, Kepler, Fermat and Napier, and the artist, Dürer.

Until fairly recently, magic squares were deemed to have magical properties. To the Jews, the third order magic square, or its cabalistic equivalent formed by using the odd numbers only (Figure 12–3), concealed the forbidden name Yahweh (Jehovah). The Jews, like the Greeks, used letters of the alphabet as numerals. The letters which represented the magic constant fifteen were the first two letters of the Hebrew for "Yahweh."

Figure 12–3 For the Jews this array concealed the forbidden name of Jehovah.

Medieval astrologers and physicians believed these arrays possessed mystical properties. For example, a magic square engraved on a single plate had the power to ward off the deadly plague. Muslim West Africa was considerably interested in the subject. Dr. Bivar, of the School of Oriental and African Studies, University of London, wrote: "Medieval scholars, perhaps hopeful of obtaining for their exercises a subsidy from the superstitious, used to represent the squares as having supernatural powers as charms and talismans. Examples of the squares reproduced in manuscripts are intentionally defective. The student was supposed to have sufficient knowledge of the theory to rework the calculations for himself and correct any errors. It seems to me possible that the squares could also have been used in connexion with cipher communication. Islamic scholars refer to these pursuits as *Ilm al-asrār*, 'The science of secrets,' and instruction in their composition was restricted to chosen pupils."

One of the scholars of Muslim West Africa was Muhammad ibn Muhammad. My introduction to his work came through an article by H. I. Gwarzo entitled, "The Theory of Chronograms as Expounded by the Eighteenth Century Katsina Astronomer-mathematician Muhammad B. Muhammad." The article is a translation of excerpts from an Arabic manuscript written in 1732 by a native of northern Nigeria. The selections deal with the use of letter symbols to represent numbers in the writing of dates. These letters are from both the Eastern (Egyptian) and the Western versions of the Arabic alphabet. Gwarzo compares them with Greek alphabetic representation of numbers and mentions their similarity to both Hebrew and Coptic. We have a similar custom of writing dates in Roman numerals in testimonials and cornerstones.

Gwarzo's description of Muhammad ibn Muhammad as an early eighteenth century astonomer, mathematician, mystic and astrologer living in Katsina (now northern Nigeria) excited my curiosity. This was my first encounter with a mathematician native to an African country south of the Sahara, and I undertook a search for more information. With the assistance of several extremely helpful scholars, I at last obtained a translation of several pages of the manuscript.

The title page begins: "Muhammad ibn Muhammad, al-Fullānı al-Kishnāwīal-Sūdāṁ." The title in Arabic is followed by the English translation, and descriptive notes: "'A Treatise on the magical use of the letters of the alphabet.' Contemporary manuscript with marginal notes in a later hand. Arabic manuscript 1732." Only pages 1–20, 91–100, 131–140, and 171–179 are in the library of the School of Oriental and African Studies, University of London. (Dr. Bivar wrote that a copy of the manuscript is in the former Khedival Library in Cairo.) I have copies of page 1 and pages 91–94.

The title of the manuscript gives no clue as to the contents of the sections in the library of the School of Oriental and African Studies. At least half the pages after page 91 contain magic square arrays, sometimes nearly filling the page. The word "magic" in this context refers to the magical use of numbers in medicine, in other words, the subject of numerology. Although the title refers to the "magical use of the letters of the alphabet," the arrays shown in the pages in my possession are written in East Arabic numerals. The author gives several formulas for the construction of magic squares having an odd number of rows and columns, and examples of each.

Page 91: Before launching into the procedures for the construction of magic squares, Muhammad ibn Muhammad warns the reader not to be misled by appearances. Things which look different may have the same meaning. Then he gives some words of encouragement. "Do not give up, for that is ignorance and not according to the rules of this art. Those who know the arts of war and killing cannot imagine the agony and pain of a practitioner of this honorable science. Like the lover, you cannot hope to achieve success without infinite perseverance."

He further admonishes his students to "work in secret and privacy. The letters are in God's safekeeping. God's power is in his names and his secrets, and if you enter his treasury you are in God's privacy, and you should not spread God's secrets indiscriminately." Muhammad was of the Fulani people, noted for their fanatic devotion to Islam.

Figure 12–4   *East Arabic numerals, used by Muhammad ibn Muhammad, the eighteenth century Muslim African scholar.*

     A detailed explanation of the construction of the 3 × 3 square follows. Keep in mind that Arabic is written from right to left.

    The 3 × 3 case is obvious, but if you do not understand it the construction is in three stages.

1.   You put 1 in the middle of the lowest row. This means it is in the second empty place. You look for its pair, but you do not find it. [Muhammad is referring to the cell diagonally down and to the right of the preceding number.] So you put 2 at the head of the column which will be the first position in that column. [Here is an example of a defective square, mentioned in Dr. Bivar's letter. In the upper right corner, where a 2 should be, the manuscript has a 6 (Fig. 12–5, top right). The East Arabic numeral 2 is a mirror image of 6.] Again, the pair of 2 does not appear, so put 3 in the last position of the second row. This finishes the first stage.

2.   Now begin the second stage. Begin to count from the position you ended with. This is the last position in the second row. Count three positions going down, so you end up with two positions below (the square). So you count up from two positions from 3. Put 4 in the top row in third position. Its pair, which is the middle, is 5. The pair of 5 is 6.

3.   Third stage: Begin with 6 and count three up. Put 7 in the second position. 7 has no pair, so put 8 in the last position of the lowest row. It has no pair, so put 9 in the highest position of the second column. You have finished (Figure 12–6).

4	9	2
3	5	7
8	1	6

Figure 12–6   Translation of third order magic square in Figure 12–5, upper right. The arrangement is identical to that in the Chinese Lo Shu.

A similar method was introduced in Europe in 1693 by De la Loubère, who learned it while he was the envoy to Siam. It applies to a magic square of any odd order. The *one* must always be placed in the center of an outer row.

The reader with less than infinite perseverance can use the following simplified procedure to construct a magic square of any odd order, based on the principles described by De la Loubère.

Imagine that the square is wrapped around a vertical cylinder, so that the first and last columns are adjoining. (We are now reading from left to right.) Now visualize the square wrapped around a horizontal cylinder. Combine the two images to represent the array in the plane as in Figure 12–7. Shade the intersecting cell of the border, and consider it

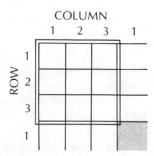

Figure 12–7   Construction of third order magic square by the diagonal method.

Figure 12–5   Magic squares appearing in Muhammad ibn Muhammad's text. For a translation of the third and fifth order squares see Figures 12–6, 12–10, 12–19 and 12–20.

occupied. A magic square of any odd order is constructed very simply. We place the *one* in the center of the bottom row and proceed by moving one step down and to the right. When we reach a block, we follow the rule of placing the next consecutive number immediately above its predecessor (Figure 12–8). By a series of ninety-degree rotations of this array, we can

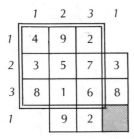

Figure 12–8   Completed   square   constructed   by the diagonal method.

obtain three more squares in which the *one* is in the center of any outer row or column (Figure 12–9).

Figure 12–9   Three other arrays can be obtained by rotating the square in Figure 12–8.

By the use of this simplified procedure and Muhammad's rules, we can reproduce his fifth order square (Figure 12–5, top center). He begins by placing *one* in the cell just below the center. When he reaches a block, at the end of each sequence of five numbers, he moves downward two cells, as he states in his instructions for the 3 × 3 case quoted above (Figure 12–10).

In this fifth order magic square the sum of each row, column, and main diagonal is 65, and the middle cell is occupied by 13, the middle number of the sequence 1 to 25. Furthermore, the sum of any two numbers that are symmetrically located with respect to the center is 26. For example,

	1	2	3	4	5	1
1	11	24	7	20	3	
2	4	12	25	8	16	4
3	17	5	13	21	9	17
4	10	18	1	14	22	10
5	23	6	19	2	15	23
1	11	24	7	20	3	

Figure 12–10   Construction of a fifth order square by the diagonal method (see Figure 12–5, top center). The magic constant is 65.

6 and 20 lie on a line through the center and are at equal distances from the center; 20 + 6 = 26, or twice 13.

Perhaps the reader will have the patience to work out Muhammad's seventh order square (Figure 12–5, top left), and check his results with the Arabic version. This one is defective, with an error in the lower right corner cell. The middle number in the array is 25, and the magic constant is $7 \times 25 = 175$.

In general, a normal magic square of odd order n consists of a square array of the integers from 1 to n^2, arranged so that the sum of each row, each column, and each of the two main diagonals is equal to the magic constant $\dfrac{n(n^2 + 1)}{2}$. By the use of the preceding formulas, we obtain squares in which the middle number in each array is $\dfrac{n^2 + 1}{2}$.

Page 92, right half of page (Figure 12–11):   The author points out that there are exactly eight different 3×3 squares, all having 5 in the center, and that all eight can be obtained from any one of them by suitable rotations and reflections.

Using as the basis the familiar third order square (Square A in Figure 12–12, and the first array left of center in the Arabic manuscript), we generate each of the other seven squares either by rotating A in the plane of the paper or by flipping it over about an axis. Assume that each cell in square A has the same number on the back as it has on the front.

A: Leave it as is (called identity operation).

B: Reflect A (flip it over) about a vertical axis.

C: Rotate A through an angle of 180° in the plane of the paper.

Figure 12–11 Text of Muhammad ibn Muhammad's manuscript, right side of page 92.

D: Reflect A about a horizontal axis.

E: Rotate A through an angle of 90° in the plane of the paper.

F: Reflect A about one diagonal as an axis.

G: Rotate A through an angle of 270° in the plane of the paper.

H: Reflect A about the other diagonal as an axis.

This set of motions, that bring a square into coincidence with itself, is termed, by modern mathematicians, the dihedral group of order 8 of the square. By the procedure above the entire set of squares is obtained directly from square A by the method of group theory. We can also derive each square from the preceding one in the sequence by reflection about a suitable axis.

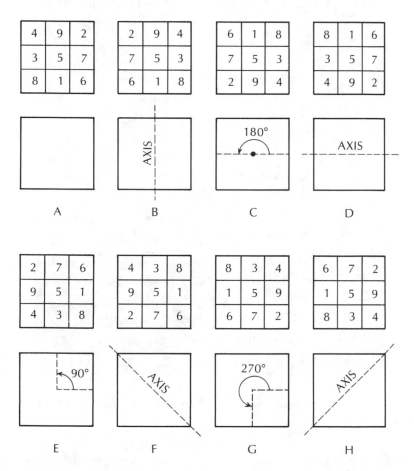

Figure 12–12 Every possible third order array can be obtained by suitable rotations or reflections about an axis of one square (see Fig. 12–11).

Figure 12–13 Text of Muhammad ibn Muhammad's manuscript, left side of page 92.

Page 92, left half of page (Figure 12–13): Here Muhammad presents two methods for constructing squares using the knight's move in chess. Again we will annex a border below and to the left of the 5 × 5 square, this time adding two rows and two columns. Each move consists of one step to the left and two steps down. Figure 12–14 shows the construction of the magic square in Figure 12–13, on the right side of the half-page. At the end of each sequence of five integers, at which point we reach a block, we place the first number of the new sequence two cells to the left of the number last used.

		13	25	7	19	1
		17	4	11	23	10
	14	21	8	20	2	14
6	18	5	12	24	6	18
	22	9	16	3	15	22
	░░░	13	25	7	19	
10	17	4		23	10	

Figure 12–14 Construction of a fifth order magic square by knight's moves, as in chess (see Figure 12–13, right side of page). The sum of the numbers in each of the five rows, five columns and ten diagonals is 65.

The knight's move array not only has the magic constant (65 in the 5 × 5 case) as the sum of each row, column, and main diagonal, it has the additional feature that any sequence of numbers in a spiral around the imaginary horizontal or vertical cylinder has this sum. For example:

$$21 + 4 + 7 + 15 + 18 = 65$$
$$15 + 24 + 8 + 17 + 1 = 65$$

The square on the left side of the manuscript half-page (Figure 12–13) is constructed similarly by knight's moves. The difference lies in the move at the conclusion of each sequence of five numbers. In this array the next consecutive integer goes into the cell just above the preceding one. Note the defect in the center cell, which should contain 15, not 5.

The reader may want to try his hand at the 7 × 7 square (Figure 12–15), constructed by knight's moves. In this array each sequence of seven numbers begins three spaces above the last cell of the preceding sequence. For example, 8 is three cells above 7, and 15 is four cells below 14 (equivalent to three cells above 14).

32	14	38	20	44	26	1
48	23	5	29	11	42	17
8	39	21	45	27	2	33
24	6	30	12	36	18	49
40	15	46	28	3	34	9
7	31	13	37	19	43	25
16	47	22	4	35	10	41

Figure 12–15 Seventh order square constructed by the knight's move method. The sum of each of the seven rows, seven columns and fourteen diagonals is 175. This square appears on page 93 of Muhammad's text.

Page 93: Here we have an amazingly simple method for writing squares of odd order (Figure 12–5, bottom). Andrews presents the procedure in a somewhat different form (page 17) as follows: To construct a 3 × 3 square, first write the integers from 1 to 9 diagonally as in Figure 12–16. Within the center square alternate cells remain empty. The numbers from outside the main square are now transferred to the empty cells on the opposite sides of the square without changing their order (Figure 12–17).

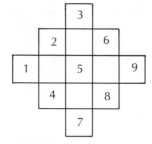

Figure 12–16 A magic square can be constructed by writing the first nine numbers in their normal sequence in a diagonal arrangement, and then transferring the numbers on the outside of the three-by-three square to empty cells on the opposite sides of the square.

2	7	6
9	5	1
4	3	8

Figure 12–17 Completed square constructed by the method described in Fig. 12–16.

Apply this procedure to Muhammad's 5 × 5 case, writing from right to left, as in Figure 12–18. The translation of Muhammad's array appears in Figure 12–19 and 12–20. The result is a mirror image of the magic square on manuscript page 91 (Figure 12–10).

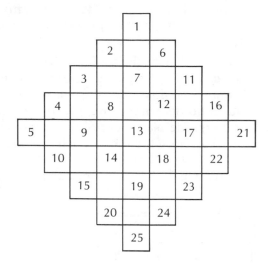

Figure 12–18 *First step in the construction of a fifth order square by the method described in Figure 12–16.*

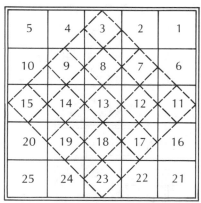

Figure 12–19 *Construction of a five-by-five magic square by the method described in Figure 12–16.*

3	20	7	24	11
16	8	25	12	4
9	21	13	5	17
22	14	1	18	10
15	2	19	6	23

Figure 12–20 *Translation of the array in Figure 12–5, lower left, constructed by the method described in Figure 12–16.*

Page 94: A novel method for the construction of odd-ordered squares by annexing suitable borders is described on page 94 of the manuscript and probably continued on subsequent pages. W. W. Rouse Ball explains this procedure on page 145 of his *Mathematical Recreations and Essays.*

We choose the simplest case, again the fifth order array. The object is to draw a 3 × 3 square within a 5 × 5.

Step 1: Fill eight (half) of the sixteen border cells with the integers from one to eight written in some order, such that none of the inner rows or columns, and neither of the main diagonals, has both end cells filled, and that no border row or column contains more than three of the first eight numbers. The placement of the numbers depends to some extent on trial and error. Since the sum of each row and each column must eventually be 65, and the largest number in the whole array is 25, clearly we cannot write 1,2, and 3 in the same row or column (Figure 12–21).

7	8			3
				2
5				
6				
		4	1	

Figure 12–21 First step in the construction of a three-by-three square within a fifth order array.

7	8	22	25	3
24	12	17	10	2
5	11	13	15	21
6	16	9	14	20
23	18	4	1	19

Figure 12–22 Completed third order array within a five-by-five magic square, described on page 94 of Muhammad ibn Muhammad's text.

Step 2: Complete the border by inserting the numbers 18 through 25 in the empty border cells, so that the sum of the two cells at the opposite ends of each of the three inner rows and columns, and of the two principal diagonals, is 26.

Step 3:   Now arrange the numbers from 9 through 17 in the center so that they form a magic square with the magic sum of 39. This can be done by adding 8 to each element of any normal 3 × 3 square. The resulting fifth order square has a magic constant of 65, the sum of 39 and 26 (Figure 12–22).

Muhammad illustrates the method by constructing a 9 × 9 and an 11 × 11 array, building them layer by layer by the border square procedure. Here one must, indeed, have the infinite perseverance of the lover!

Nowhere in these four pages is there a reference to the application of these squares to magic, or to the use of the letters of the alphabet, as the translation of the title of the manuscript leads us to expect.

Whether Muhammad ibn Muhammad made original contributions to number theory or any other branch of mathematics, I cannot say. Several of his manuscripts are in the former Khedival Library in Cairo. He had a reputation as a teacher of astrology and numerology; "he acquired this learning in his own country," reported an Egyptian colleague. In 1730 he made a pilgrimage to Mecca, the dream of every devout Muslim, and subsequently he spent time in Cairo, where he died in 1741. Muhammad's teacher was Muhammad Alwālī of Bagirmi, a scholar of note, and the author of verses on divination and astrology, wrote Muhammad Bello, the nineteenth century Sultan of Sokoto. His works suffered the fate of most African manuscripts—they have not survived.

SECTION 5
PATTERN AND SHAPE

The strong African sensitivity to form is evident in the construction of houses, in the carving and embellishment of masks and sculptured figures, and in the decoration of simple domestic objects.

   The circular house is the architectural mode most common to traditional societies and the one most admirably suited to the prevailing level of technology. However, in the forested regions, where timber is readily available, and in areas of Muslim or European influence, the rectangular dwelling is popular. In recent times the possession of a rectangular home has become a status symbol in such widely separated parts of Africa as Ghana and Kenya.

   African art forms are intimately related to religious, social and domestic customs that are alien to the experience and traditions of Europeans. It is beyond the scope of this book to discuss the religious, ceremonial, or utilitarian purposes for which the art objects have been created. My approach to the vast and complex field of African art is purely from the point of view of an analysis of forms and patterns.

Figure 13–1 Kibo Art Gallery, on the slopes of Mt. Kibo, one of the peaks of Mt. Kilimanjaro. The entrance, on the right, is an immense thatched structure in the style of the Chagga beehive house (see Fig. 13–4).

CHAPTER 13

GEOMETRIC FORM IN ARCHITECTURE

The Round House

> The round warm hut
> Proud to the last
> Of her noble sons
> And daughters
> Stands besieged.

The author mourns the passing of the familiar furnishings of the round house, and concludes:

> All this and much more,
> Slowly and slowly disappears:
> Slowly and slowly iron appears,
> Lays a seige on the roof
> And takes prisoner the pot and the gourd,
> The plate, the cup, the lamp,
> What's this but a change
> To the new oblong house?
> The round mud hut is no more.

The lament of the Kenya poet Joseph Waiguru describes the passing of the old forms, the traditional way of life. Today one sees mainly square and oblong houses with peaked roofs in the newly settled lands of Kenya. Joseph Wambua was showing us the *shambas* (farms) of his parents and uncles. He and his brother had built a sturdy home of sun-baked bricks which they had molded themselves. Then he pointed into the distance at a small round structure in the midst of several larger, rectangular buildings. "That is the old kind of Kamba house. Now they use it only for storage."

The African adapts his home admirably to his means of subsistence, to the available materials, and to the requirements of the climate. The circular house in its many versions is found throughout the continent. The circle, of all closed geometric shapes in the plane, encompasses the greatest area within a given perimeter. Confronted by a scarcity of building materials and by the urgency to erect a shelter without professional assistance and in the shortest possible time, the African chooses the circle as the most economical form. He is not unique; round houses are constructed in the Arctic as well as at the Equator.

Let us compare the areas of a circle, a square and a rectangle of a given perimeter (Figure 13–2). A circle with a diameter of fourteen feet has a circumference of approximately 44 feet, close to the dimensions of the base of the Kikuyu home. A square of equivalent perimeter measures eleven feet on the side; a seven-by-fifteen foot rectangle also has a perimeter of forty-four feet. Now compute the areas:

$$\begin{aligned}
\text{Circle:} \quad & 22/7 \times 7^2 = 154 \text{ square feet} \\
\text{Square:} \quad & 11^2 = 121 \text{ square feet} \\
\text{Rectangle:} \quad & 7 \times 15 = 105 \text{ square feet}
\end{aligned}$$

For a given perimeter of 44 feet, the area of the square is smaller than the area of the circle by 33 square feet, or 21%. With the rectangle we lose 49 square feet, or 32%, compared with the circle. Obviously the round house has the lowest requirements in terms of materials and expenditure of time and energy.

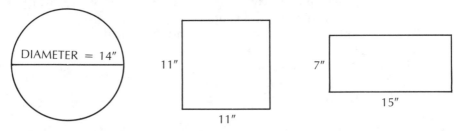

Figure 13–2

Labelle Prussin has studied the architecture of West Africa, and has this to say: "I did a simplified 'engineering analysis' comparing the round and square houses in the savannah, with equal volume, for bending, shear, bearing, etc., given similar vertical and horizontal loading, assuming constants in humidity, temperature, and soils, vegetation and technology. Of course, the round house won by a landslide."

In its simplest form the circular house is well suited to the needs of nomadic people, the hunters and the cattle herders. Even in agricultural societies, where rotation of land is practised, houses may be abandoned after just a few years.

A simple type of circular structure is built by the Ituri Forest Pygmies, or Mbuti, in the northeast Congo (now Zaire). Colin Turnbull describes this construction in *The Forest People*. Masemba first built the founda-

Figure 13–3   Fipa hemispheric thatched house (southern highlands, Tanzania). Village Museum, Dar es Salaam.

tion, driving the saplings into the ground. Then she stood up and bent the tops together over her head, twining smaller saplings across to form a dome-shaped lattice framework. Large heart-shaped leaves, almost two feet across, were hung like tiles, overlapping each other and forming a waterproof covering. Once the leaves have settled down, not even the hardest rainfall can enter the house, until the leaves begin to curl with age. A roof cover of dead branches helps to hold the leaves flat.

In a new hut the rain may send a waterfall to the floor, and it is the wife's job to get up to repair the leak. In fact, when a man wants to go to sleep in peace, he tells his wife, "Get up and fix the roof!" She gets the message.

Some Fulani cattle herders of West Africa construct round temporary huts of corn stalks, tied together at the top, which last until it is time to move on to greener pastures. The inhabitants of Lake Kyoga, in Uganda, build their homes of papyrus stems, thatched to resemble inverted baskets ten to fifteen feet high. The floating dwellings can be transported from place to place by canoe, as their owners search for plentiful supplies of fish.

When a Chagga built his traditional beehive-shaped house (Figure 13–4) on the fertile slopes of Mount Kilimanjaro, he called upon the tallest man he knew. This neighbor would lie flat upon the site of the prospective home, with his arms outstretched. The span from the fingertips of one hand to those of the other is called a *laa*. To mark off the circumference, the builder tied a hoe to a rope of length equal to the desired radius, two to three *laa*. The rope was attached to a peg, and as he walked around this peg, he drew a circle with his hoe. The door height was equal to the span of the man's arms; its width was the circumference of his head, measured by a string.

Figure 13–4 The author in front of a beehive-shaped Chagga house (Mt. Kilimanjaro). Village Museum, Dar es Salaam.

Today the Chagga people, prosperous from the proceeds of their coffee cultivation, build roomy oblong houses. But the traditional beehive shape is preserved in the immense entrance hall to the Kibo Art Gallery, supported in the center by a massive tree trunk (Figure 13–1).

In a sedentary agricultural society, the round house often takes the form of a cylindrical structure topped by a cone-shaped roof. The walls

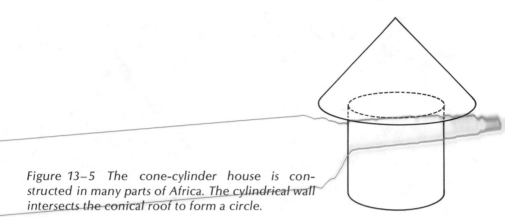

Figure 13–5 The cone-cylinder house is constructed in many parts of Africa. The cylindrical wall intersects the conical roof to form a circle.

are of mud, and the conical roof, gently sloping or rising to a sharp point, is thatched with local grasses. The traditional house of the Kikuyu of Kenya is a strong, comfortable, wellbuilt structure, admirably adapted to the requirements of its users. It consists of a cone-shaped roof of thatch, set upon a cylindrical base (Figure 13–6). A circle, about fifteen feet in diameter, is marked out on the ground, and nineteen holes are dug at equal distances from one another, allowing a wider space for the door. In each hole is placed a five-foot post. Then a three foot by four-and-a-half foot rectangle is marked off in the center, and a five-foot post set up at each corner as an additional roof support. The upper ends of all the outer posts are split, so that a ring of strong bark strips can be placed on the circle of posts. The wall is formed by a network of wattling fastened to the posts, and then daubed with clay. Frequently a coat of whitewash is applied to the smooth wall, and the house is a pleasant sight, gleaming in the equatorial sun.

The roof consists of a center axis to which are fastened pliant rods. These are lashed to a circular hoop, and then forced to assume a convex shape. A larger circular hoop reinforces the frame. The thatched roof projects several feet over the cylindrical base, to form a shady verandah and to carry off the rain.

With the assurance of stability, people build their homes for permanence and beauty. Some houses have murals painted on the inner or outer walls, others feature low-relief ornamentation around the doorways or decorative treatment of the roof. Frank Willett describes the architectural technology of the Ham (or Jaba) around Nok, in northern Nigeria (page 116): "[Their homes] afford an outstanding example of the exploitation of the possibilities of mud in architectural design. . . . The Ham have

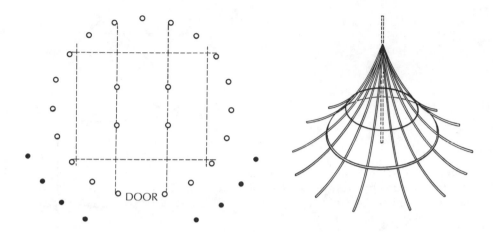

Figure 13–6   Construction of a Kikuyu (Kenya) cone-cylinder dwelling: (left) floor plan, and (right) roof support showing vertical axial stake. (After Routledge.)

Figure 13 – 7   Inside view of the conical roof of a Nyakyusa house (southern highlands, Tanzania). Village Museum, Dar es Salaam.

Figure 13–8   Nyamwezi cylindrical house with a cone-shaped roof (west central Tanzania). Village Museum, Dar es Salaam.

a relatively simple technology, yet they have fully grasped the plastic potential of mud architecture in a way which surpasses that of medieval and later European builders in mud and thatch.... The Ham have produced a house design which is a veritable sculpture for living in, something more than the mere machine for living in which Le Corbusier demanded."

A remarkable feature of African architecture is the over-all consistency in form. In many societies the houses are arranged in compounds, with separate houses for each of the several wives (when the husband can afford more than one), a storage shed for each wife, and dwellings for married sons or daughters. Round houses are generally arranged in round or elliptical compounds, rectangular houses in oblong compounds.

Significance of the Round versus the Rectangular House

In his novel, *The River Between*, James Ngugi describes the Kikuyu village—scattered over the ridge were round mud-walled thatched huts, standing in groups of three or four. "Joshua's house was different. His was a tin-roofed rectangular building standing quite distinctly by itself on the ridge. The tin roof was already decaying and let in rain freely, so on top of the roof could be seen little scraps of sacking that covered the very bad parts" (page 32). Joshua was the first convert to Christianity in his village, and his rectangular house unmistakably set him apart.

One of the architectural styles analyzed by Labelle Prussin is that of the Konkomba of northern Ghana, a people who rejected Islam and all other external cultural influences. Their houses are cylindrical, of mud covered with lime, and topped by a thatched conical roof. Typically, their houses are arranged in oval compounds (Figure 13–9). Above the opening into each room three or four cowrie shells are imbedded in the mud, three symbolizing male, and four, female, to indicate the sex of the room's occupant. "One of the two rooms in the [young men's] subcompound, built in the strikingly rare rectangular form, belongs to a young, unmarried man who is a schoolteacher. To become a teacher is an achievement of no small account in Konkombaland, and the rectangular room is a symbol of the prestige he enjoys" (1969, page 46). In a personal communication, Mrs. Prussin adds, "The presence of rectangular houses in circular compounds, as well as the status assigned to them, is a widespread phenomenon in the savannahland of West Africa."

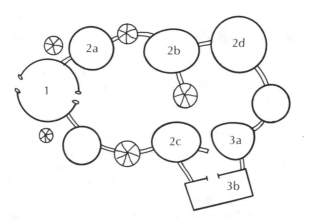

Figure 13–9 Compound in a Konkomba village (northeast Ghana). 1. antechamber; 2. wives' rooms; 3. unmarried sons' rooms. The rectangular room is a symbol of the high status of its occupant, an unmarried son who has become a schoolteacher. (After Prussin: Architecture in Northern Ghana.*)*

Figure 13–10   Rodah Zarembka, her mother and sisters are winnowing beans near Rodah's "house with corners," Kamba, Kenya.

To the older people of a society this transition from round to square or oblong is more than just an innovation in geometry—it represents the passing of the old order, the loss of security, as illustrated by this anecdote. After Kenya had achieved independence and many of the European settlers had departed, the government divided their former holdings into small plots. Rodah and David Zarembka purchased a six-acre plot on which they built several oblong houses for themselves, some of Rodah's family, and the livestock. Proud of her *shamba* (farm), Rodah invited her grandparents to stay with them for a while. "Do you live in a house with corners?" asked the grandmother. Told that they did, the grandmother replied, "No, we can't live in a house with corners. There is no pole to support the house. And besides, we would get lost in such a house!" The old folks were accustomed to the traditional round dwelling, similar to that of the Chagga, constructed on a framework of pliant young trees stuck into the ground in a circular arrangement. In the center stands a thick post, the support of the structure. The saplings are tied together at the top, and a framework of parallel concentric circles is placed upon this foundation. Then the whole thing is thatched with grasses.

Figure 13–11   Roof ready for thatching, northeastern Tanzania.

The traditional home of the Zulu is an umbrella of thatch built on a woven circular framework of light branches. Forced into ugly shacks or barracks-type buildings in South African towns, the Zulu reverts to the familiar circle in his religious practices, typical of the many small sects that combine Christianity and traditional Zulu beliefs. The congregations may meet indoors or out, but the circle is always painted on the concrete floor or drawn in the mud. "The importance of the circle can only be grasped when one considers the round world in which the Zulu traditionally lived. . . . Amidst the harsh rectangularity, the gridiron of Western urban life of which the Zulu cannot become a part, their circles recapture this round world" (Fernandez, page 48).

Some people gave up the "round mud hut" in favor of the oblong house; others adapted it to modern conditions. The Tswana of southern Africa have put up new round buildings of brick or stone, with thatched roofs. In the Botswana mining town of Selebi-Pikwe, temporary houses of iron sheets have been constructed in the shape of the octagon, probably the closest approximation to the circle permissible with the available materials.

The Rectangular House

Early in the twentieth century the German anthropologist Leo Frobenius visited many African lands. He was one of the few sympathetic European observers of his time, and he returned to write glowing accounts of the cultures he had encountered. He was particularly impressed by the homes of the Yoruba people of southwest Nigeria—rectangular apartments, with walls of clay, and raftered ceilings covered with clay, set in rectangular compounds, each surrounded by a great wall. A gabled roof of poles and foliage covered each apartment and dropped far beyond the walls to form a verandah. In the center courtyard large pots caught the rain as it dripped off the eaves, a construction similar to the "impluvium" of the ancient Mediterranean societies.

In the forested regions of southern Nigeria and the Guinea Coast, are villages and towns of rectangular homes set in compounds. The walls are often painted with murals or decorated with relief patterns, while carved wooden doors and posts adorn the homes of the wealthy and the privileged. Igbo women like to paint the outer walls of their red-earth houses in white, green, red and black geometric patterns. In the Congo and Zambezi regions, one finds rectangular houses hung with beautifully woven mats in intricate patterns, or decorated with murals.

The *tembe* of the Hehe people of Tanzania is a large oblong structure of many rooms surrounding an inner courtyard. The mud walls are ten inches thick, and the thatched flat roof is daubed with earth. "It's as cool as an air-conditioned house," declared a Tanzanian friend.

A unique dwelling is constructed by the nomadic cattle-herding Maasai of Kenya and northern Tanzania. The village, or *manyatta*, consists of a series of low, rectangular round-topped houses made of supple stems,

Figure 13–12   Rectangular courtyard of a home in Ghana.

Figure 13–13 Maasai dwelling, constructed of supple stems plastered with mud and cowdung. These houses are placed end to end to form a large circular enclosure, about forty yards across, in which the herds of cattle, sheep and goats are protected during the night. Village Museum, Dar es Salaam.

and plastered with mud and cow dung. The shape of the structure resembles that of the covered wagon, famous in the "go west" era of the United States (Figure 13–13). These units are placed end to end to form an enclosure in which the herds of hundreds of cattle, sheep, and goats are safe during the night.

The introduction of rectangularity in the architecture of the West African grasslands and the East Coast has been attributed to the influence of Islam. In 1495 Askia Muhammad the Great, most famous ruler of Songhai, employed guilds of masons to construct cities of oblong buildings—mosques, Koranic schools, and ordinary dwellings. The French traveler Caillé reported in the early nineteenth century that as he traveled northward in Bambara country (now in modern Mali), he observed more and more rectangular shaped buildings. On the other hand, not all Islamized folk gave up their traditional circular houses. Today one may find both the circle and the square in the same village, and even in the same compound.

Figure 13–14 Typical house facade at Jenne, Mali, illustrating the synthesis between elements of indigenous architecture and elements introduced from North Africa via trans-Saharan contact and trade. Courtesy of Labelle Prussin.

Outstanding Buildings

When one thinks of European architecture, images of Gothic cathedrals and medieval castles come to mind. Africa, too, has its distinctive and splendid palaces and religious shrines.

The pyramids and columned temples of ancient Egypt still stand today, a memorial to the wealth and power of the divine kings and the priests of ancient Egypt, and a tribute to Egyptian architects and craftsmen. Less familiar are the pyramids and columned temples of the kingdom of Kush (now northern Sudan), which flourished two thousand years ago. The houses in its capital city, Meroë, consisted of a series of rooms all opening upon a central courtyard, a style still prevalent there and in other regions of Africa. In contrast, people in the countryside constructed reed huts in the shape of elongated beehives, gathered and tied at the top; our information for this model comes from a design engraved on an ancient bronze bowl. South of Meroë, near Khartoum, is the "Great Enclosure" at Musawwarat, an extensive multistructural building unique in the Nile Valley. A huge water basin served both as a reservoir for drinking water and for the irrigation of the surrounding fields. The columned Lion Temple

nearby reveals a sculptural representation of both man and beast that is entirely different from that of Egypt.

In more recent times, those African societies which had kings generally housed them in more imposing versions of the prevailing style of architecture. The palaces of the rulers of Buganda differed from the more humble homes in their size and decoration; both were in the same beehive style and constructed of the same materials. The residence of King Munsa of the Mangbetu, in the northern Congo, was a large airy hall about 150 feet long, 75 feet wide, and 50 feet high. Despite its size, it was built in the same style as the homes of his subjects.

Frobenius remarked about the remains of the ancient palace of the Oni of Ife, the traditional home of Yoruba culture, that it resembled more recent Yoruba architecture. In earlier centuries European travelers were amazed when they entered African cities like Benin, or the Asante capital of Kumasi, to find impressive architecture, well laid streets, and a cleanliness rare in their European towns. Kumasi was called a "garden city." In the early nineteenth century, the wealthy lived in beautiful homes that were built around an internal courtyard and equipped with internal sanitary facilities.

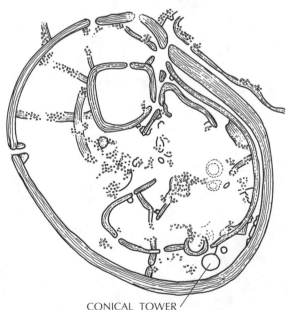

CONICAL TOWER

Figure 13–15 Ground plan of Zimbabwe's Great Temple, one of the most spectacular ruins in Africa. The earliest structure was erected over a thousand years ago. The conical tower appears to be of solid masonry; its original function is still shrouded in mystery.

Figure 13–16   The Temple at Zimbabwe, seen from the air. The conical tower, thirty-four feet in height, is in the foreground. Courtesy of David Attenborough.

The wealthiest kingdom of south central Africa was Monomotapa, location of the fabulous gold mines. The greatest cultural remains of the kingdom are the celebrated ruins of Zimbabwe. The Temple has a massive outer wall thirty feet high and twenty feet thick, built in the shape of an ellipse two hundred feet wide and three hundred feet long. There is also an inner wall and a solid conical tower like nothing else in the world. Another group of ruins is called the Acropolis, and between the two are the Valley Ruins. All the construction is in dry stone—no mortar is used to hold the stones together. The walls consist of a roughly dressed stone facing with a rubble core. The finest construction is evident in the elliptical Great Wall and the thirty-four foot Conical Tower, where the stones are precisely cut and laid in neat courses.

There are numerous monoliths, soapstone carvings, centuries-old workings of gold mines, and a wide variety of precious objects. Unfortunately, the area was ransacked by European treasure hunters, who destroyed most of the evidence needed for archaeological investigations.

Figure 13–17 The Sankore Mosque at Timbuktu was constructed in the early fourteenth century, during the reign of Mansa Musa of Mali. Although built of mud, it has endured through the centuries because the permanent wooden scaffolding, like bristles projecting from the walls, has provided a method for continuous repair of the building. Courtesy of Labelle Prussin.

Ruins of similar stone buildings are scattered throughout modern Zimbabwe (Rhodesia). Frank Willett writes (page 136):

Zimbabwe is by no means the only group of stone buildings in southern Africa. The Khami Ruins are hardly less impressive, and Naletale in Rhodesia, which was occupied by the BaRozwi in the seventeenth and eighteenth centuries, has outstandingly fine decorative walling.

All are complexes of circular buildings, probably deriving their form from the more humble cylindrical mud dwelling.

The massive pyramidal forms of the Sankore Mosque at Timbuktu, dating from the early fourteenth century, illustrate the African architectural style as modified by Islamic influence. Although it is constructed of mud, it has endured throughout the centuries because of the scaffolding, like wooden bristles, incorporated into the walls, which allows continuous repair of the building.

Another architectural triumph is the construction in the thirteenth century of ten Christian churches carved from solid rock, under the aegis of King Lalibela of Ethiopia. There are no records to indicate the techniques used to hollow out the vaults and construct the columns. The churches are beautifully decorated with geometric carvings and bright murals. Mathematical calculations must have been involved, since a slight miscalculation in the measurements would have ruined the whole structure.

Thus both Islam and Christianity produced styles in African architecture that are truly unique.

PART I GEOMETRIC FORM AND
PATTERN IN ART

In Africa Art is Everywhere

Art is part of the moving continuum of daily and ritual existence in Africa; it is widely diffused throughout society and the life of the individual. The African does not distinguish between fine and applied art. In religious ceremonies, elaborate costumes and carved wood masks and figures are combined with music and dance to involve the whole community in a unified artistic expression. Even in areas of Islamic influence, the Muslim ban on the representation of human and animal figures has not eradicated traditional art forms. The articles of every day life—woven cloth, calabashes, ordinary cooking spoons—all are decorated, usually with patterns having symbolic meaning. Among many peoples, the graphic arts play a role

Figure 14–1 Baskets used as stands for milk pots, Uganda. British Museum.

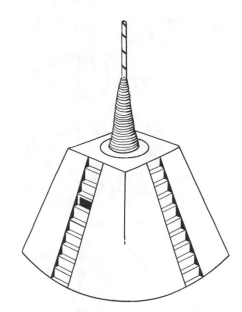

Figure 14–2   Model of the universe in the mythology of the Dogon of Mali. The circular base is the sun and the square top, the sky, with a circle for the moon. It is called the Granary of the Master of Pure Earth. (From Griaule: Conversations with Ogo-temmêli)

similar to that of writing in conveying ideas and recording history.

Traditions in sculpture go back many centuries, and find their expression in masks for religious and social ceremonies, in representations of deities and spirits in human, animal, or abstract form, in figures of special significance for certain occasions or for the home of a king or chief. The African artist is not afraid to distort the natural figure to achieve emphasis, strength, and force. On the other hand, he is capable of producing completely naturalistic sculpture, as in the Yoruba art of Ife, dating back almost a thousand years. African art is characterized by conventionality in style that is based on long-established precedent and expressed in a variety of forms.

Geometric Form

As their model of the universe, the Dogon of Mali picture a structure having the sun as a circular base and steps leading up to a square top, the sky, enfolding a circular moon (Figure 14–2). The sun is represented by a circle in the Asante gold weights. The circular faces of the *akua mma* (Figure 14–3), the dolls carried by expectant mothers, express the Asante ideal of beauty.

The tendency to distort natural forms for the purpose of stressing certain characteristics, and the wide application of symbolic motifs in

Figure 14–3 Akua ba *doll of the Asante, characterized by a round flat face, long geometrically stylized neck, and conical torso. It is carried by the expectant mother to insure the health of her child. Private Collection of Mr. and Mrs. M. Abelson, Cambridge, Mass.*

decoration, lead to emphasis on the geometric aspect in African art. European cubism of the early twentieth century took its inspiration from the sculpture of Africa. As just one example, the concave faces of Picasso's "Les Demoiselles d'Avignon" resemble the masks of the Kwele, Kota and other peoples of the Congo area (Figures (14–4, 14–16, 14–17)).

If one wanted to survey the whole field of geometric design in Africa, one would have to catalog almost every aspect of life, from commerce to courtship. In Figure 14–6 are several of the delicately fashioned Asante brass weights used for weighing gold dust; their geometric motifs embody words of wisdom. The decorated Yoruba calabash (Figure 8–4) and Kamba gourd (Figure 8–3) have a recognized spiritual significance as well as material use in the market place. As for love, the Zulu maiden sends her lover messages in the form of beaded bands incorporating conventional symbols of affection.

Figure 14–4 Kwele mask, Congo Republic. The heart-shaped face is characteristic of this region.

Figure 14–5 The Western Ijo and the Urhobo, Nigeria, represent the head of
the household making a sacrifice to the family guardian spirit, depicted as a
fantastic animal. Note how the planes of the sculpture meet in sharp angles.
Museum of Primitive Art, New York.

Figure 14–6   Asante brass weights in geometric shapes, used for measuring gold dust currency. Many of the designs have symbolic meaning; for example, a zigzag line represents the fire of the sun. British Museum.

Repetitive Patterns

The principal item of feminine attire in parts of East Africa is the *khange*, the brightly colored rectangle that a woman wraps about her body, drapes over her head, and uses as a sling in which to carry her baby on her back as she goes about her tasks. In addition to the decorative design on the body of the cloth, each rectangle has an all-around border in a geometric pattern representing a bowl, a cashew nut, a star, or a sweet potato, to name but a few. These conventional designs, stylized versions of the objects they represent, have been handed down from generation to generation.

A large number of motifs in Kuba art are based on interlacing patterns, many derived from weaving techniques and the knotting of string for fishnets. Another source is the game children like to play in the sand (see Chapter 10). The object of this game is to draw certain traditional patterns in one continuous line without raising the finger from the surface. Inevitably certain parts of the line are eradicated when crossed by later portions.

The most common motif is *Mbolo*, a design derived from interlacing two pairs of strands. It is also found in children's sand-drawing games. The pattern may cover the plane (Figure 14–7), or be continued on a long band, as in Figure 14–8,#8. The variation in Figure 14–8,#9 is named *Mgongo*, the knee.

Another motif, called *Namba*, the knot, arises from twisting two cords together (Figure 14–7). If the twists are accentuated into a sharp angle, the resulting pattern is *Nemo Kanya*, the fingers of Kanya, illustrated in Figure 14–8, #11. A similar pattern, when used in woodcarving, is called *Buina* due to its resemblance to the curved blade of a knife. Apparently the Kuba could not agree among themselves as to the names and symbolism of the designs. The sharpest disagreements occurred between men and women, since each sex interpreted the motif in the light of its own experience. Embroidery is done by women, while carving is a male prerogative. The very same pattern may have several different meanings when it is applied to more than one medium.

Turning to the Yoruba of southwest Nigeria, we find again that braiding or knotting of strands of cord inspired the patterns in both the royal beaded boots in Figure 17–1 and the carved calabash in Figure 8–4. Similar motifs appear in the wood and ivory carvings and the pottery of the city of Benin, in southern Nigeria.

The repetitive patterns that can be reproduced on a strip and on a plane surface have been analyzed mathematically. The reader is referred to Part II of this chapter, written by Dr. D. W. Crowe, for a discussion of the twenty-four different types, and their occurrence in the art of the Kuba,

Figure 14–7 Embroidered raffia cloths, Kuba, Congo (Zaire). These eighteenth century cloths are examples of several traditional patterns. The mbolo pattern, based on weaving, is represented by the cloths in the upper right, left center and right center. Namba, the making of knots, is illustrated by the cloth in the lower left. A combination of the two patterns is shown in the lower right. British Museum.

Figure 14–8   Embroidered cloths, Kuba, Congo (Zaire). British Museum. They are classified on the basis of symmetries on the strip as follows: 1. Type 5   2. Type 4   3. Plane pattern   4. Type 4   5. Type 7   6. Type 7   7. Type 2   8. Type 2   9. Type 2   10. Type 4   11. Type 1 (or 4)   12. Plane pattern.

Yoruba and Benin peoples. Here I shall discuss the seven varieties of symmetrical patterns on a strip.

Patterns on a Strip

There are many examples in African art of patterned strips, sometimes employed to separate one major section of a design from another, as in the Yoruba carved door (Figure 14–10), or as borders on cloth or carved objects, or even as decorative motifs in themselves. More often than not the particular pattern has a name and some significance in the lives of the people, as in the embroidered cloths of the Kuba people (Figure 14–8).

From the mathematical point of view, there are seven distinct methods of repeating a given pattern on a strip or ribbon of infinite length. They can be represented by repetitions of upper case letters, as shown by the mathematician, H. S. M. Coxeter, in Part II of this chapter.

The following representation uses lower case letters:

Type (1)	. . . bbbb . . .
Type (2)	. . . bᏢbᏢ . . .
Type (3)	. . . bdbd . . .
Type (4)	. . . b b . . . q q
	. . . bbbb . . . qqqq
	. . . bqbq . . .
Type (5)	. . . bdᏢqbdᏢq . . .
Type (6)	. . . bbbb . . . ᏢᏢᏢᏢ
Type (7)	. . . bdbd . . . pqpq

Figure 14–9   Raffia pile cloth, Kuba, Congo (Zaire). King Shamba Bolongongo is said to have introduced the technique of weaving this beautiful cloth in the early seventeenth century. The pattern of the outer bands is Type 4 symmetry on the strip, with a section missing. The center band is Type 3. Smithsonian Institution, Washington, D.C.

Figure 14–10 Carved wood door panel, eastern Yoruba, Nigeria. The small bands of pattern separating the rows of processional figures are, from top to bottom, Types 6, 4, 7, 6, 7 and 6. The decorations on such architectural features as door panels frequently have historical significance, as well as humor. The style is in the tradition of many centuries, and is still popular today, applied to current subject matter. British Museum.

Figure 14–11 Carved wooden bowl, Kuba, Congo (Zaire), collected in 1899. The bowl is decorated with a Type 4 symmetry pattern. Smithsonian Institution, Washington, D.C.

All the patterns are obtained from combinations of two geometric operations, translation and reflection on an axis.

To illustrate, let us make a stencil by cutting a pattern in a rectangle:

Figure 14–12

The pattern can then be repeated by various combinations of the following two operations as in Figure 14–13:

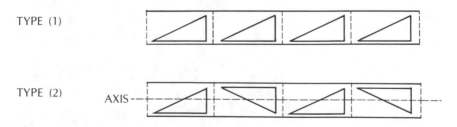

TYPE (1)

TYPE (2)   AXIS

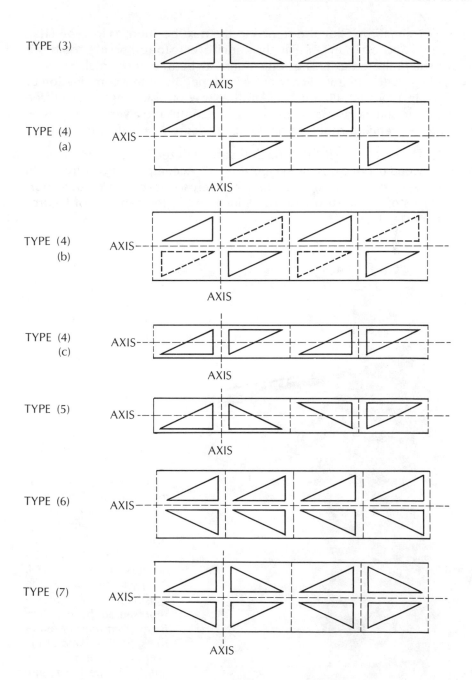

Figure 14–13 Symmetry patterns on the strip.

Translation. Slide the stencil to the next position, as in Type (1).

Reflection in an axis. Flip the stencil over to the other side, rotating it about the indicated axis. The reflection may be in a horizontal axis, as in Types (2) and (6); in a vertical axis, as in Type (3); or in a combination of both, as in Types (4), (5) and (7). Another way of obtaining the last three types, the result of reflection in both a horizontal and a vertical axis, is by turning the stencil through an angle of 180 degrees, a half turn.

The patterns in Kuba raffia cloths are analyzed in Figures 14–8, 14–9, and the wood carving design in Figure 14–11. Accompanying Figure 14–10 is a discussion of the bands separating the lively scenes on the Yoruba door. An analysis of Benin strip patterns is included in the captions of Figures 14–14 and 14–15.

Figure 14–14 The costume of the Benin flute player illustrates several types of symmetry on the strip. The bands on the hat are Type 6; the vertical borders are Types 3 and 4. A Type 5 strip is part of the waistband, and Type 7 is represented by the third band from the bottom of the skirt. The band at the bottom is another example of Type 4. Sixteenth century or earlier. British Museum.

Figure 14–15   Bronze plaque in the royal palace of Benin, Nigeria. The figure at the left displays three types of symmetry patterns in alternating arrangement. Starting at the top, they are Types 3, 7, 3 again, and 4. British Museum.

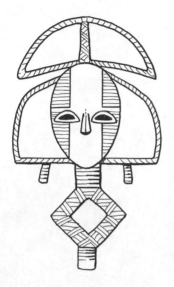

Figure 14–16   Kota funerary figure. The wooden figure is covered with brass or copper sheeting, and placed over a package containing some bones of an outstanding ancestor. The eyes and nose are formed on the inside of the concave face.

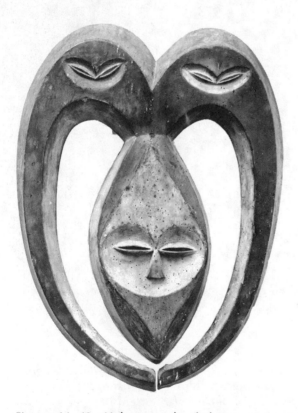

Figure 14–17   The emphasis is on concavity in this Kwele mask, Congo Republic. In some African cultures the human face is formed as though it were on the inside of an eggshell. Typical are the Kwele masks and the Kota brass-plated figures (Figure 14–16). In carving the mask, the artist sets off the face from the lateral parts of the head by a sharp edge, and paints this edge in a contrasting color. Museum of Primitive Art, New York.

Figure 14–18   Helmet mask of the men's society, Makonde, Mozambique. Many Africans scarify their faces and bodies with geometric patterns, either as identification with their clan, as an indication of social rank, or purely for esthetic reasons. British Museum.

Figure 14–19 Zulu wood sculpture, remarkable for its originality of design. British Museum.

Figure 14–20 Kikuyu wooden shields, Kenya. The patterns painted on the front of the shield (center) denote the owner's clan and age set. The reverse side (left) is black and white. On the right is a dancing board. British Museum.

CHAPTER 14
PART II GEOMETRIC SYMMETRIES
IN AFRICAN ART
D. W. CROWE

1. Introduction

A geometric analysis of the symmetries of repeated patterns, such as wall-paper patterns, shows that there are only 24 different types of patterns which can be used to cover a plane surface. Of these 24, seven admit translations in only one direction and are called *strip patterns*. The remaining 17, called *plane patterns*, admit translations in more than one direction, as well as rotations of 360°, 180°, 120°, 90° or 60°. (No other rotations, except multiples of these, are possible in a pattern whose infinite repetition covers the plane.) These 24 types were originally studied in connection with crystallography, and indeed the proof that there are only 24 possibilities is due to the crystal-lographers (notably Federov in 1891). However, it is appropriate to apply them to the study of man-made patterns, as well as to crystal structures.

The 7 strip patterns have been represented neatly by Coxeter (*Introduction to Geometry*, p. 48) as follows:

(1)	. . . LLLL . . .
(2)	. . . LΓLΓ . . .
(3)	. . . VVVV . . .
(4)	. . . NNNN . . .
(5)	. . . VΛVΛ . . .
(6)	. . . DDDD . . .
(7)	. . . HHHH . . .

We will refer to them by these numbers. Note that the only symmetries of (1) are translations along the strip. In (2), however, there are not only trans-lations (from one L to the next) but also glide-reflections, that is a glide (i.e. a translation) from L to Γ followed by a reflection in the horizontal axis. A similar analysis distinguishes each of the seven from each of the others. For example (3) admits reflections in vertical axes, while (4), (5), and (7) all admit half-turns (which interchange the two ends of the strip). A detailed analysis of the 17 plane patterns would take too long. The interested reader is re-ferred to H. S. M. Coxeter and W. O. J. Moser, *Generators and Relations for Discrete Groups* (Section 4.5) or A. Speiser, *Die Theorie der Gruppen von*

Endlicher Ordnung, for a derivation, and illustration, of the 17 patterns. We will use the standard crystallographic notation used by Coxeter and Moser.

It has long been known that all 24 plane patterns occur in Islamic art, and we ask ourselves to what extent these patterns occur in African art. In fact, of course, we find them everywhere. They appear, for example, on dyed cloths (such as the Ashanti *adinkira* or Mali *bokolanfini,* to name just two), carved calabashes, wood carvings of all kinds, as painted or mud-sculptured house decorations, in beadwork, woven mats, and so on. In the next paragraphs we will discuss just three of these, Bakuba art in general, Benin bronze castings, and Yoruba adire cloth.

2. Bakuba Art

Possibly the most striking African decorative patterns appear on the raffia pile cloths, called "Kasai velvet," of the Bakuba people of the Congo. The same characteristic patterns found on these pile cloths also decorate many other Bakuba objects, such as wall hangings, carved boxes, cups, drums,

Figure 14–21

and royal statues, and the solidified camwood powder (called *ngula*) used for ceremonial purposes. All of the 7 possible strip patterns are found on cloths and on wood carvings (except possibly type (3) on wood carvings). Of the 17 plane patterns at least 12 occur.

Typical Bakuba designs are shown on the two cups of Figure 14–21. (To consider these as plane designs, imagine the cups to be cylinders and then unroll them.) The left design, of crystallographic type pgg, admits rotations by 180° as well as glide reflections. The design on the right, of type p2, also admits rotations of 180°, but no reflections or glide reflections. An alternative arrangement of the elements of these designs, as in Figure 14–22, admits rotations of 90°. However, this third design (of type p4) never occurs in Bakuba art. This fact is especially remarkable in view of the fact that it is the same type of design as that obtained by repeated swastikas, a design which indeed occurs among near neighbors of the Bakuba, the Bayombe, whose patterns in general seem very similar to those of the Bakuba.

3. *Benin Bronzes*

When the magnificent bronze castings and ivory carvings from Benin were first brought back to Europe after the British invasion of 1897 it was assumed that this art could not be of African origin. Since some of the bronzes actually represented Europeans and European weapons, it was sometimes supposed that the art had been learned from the Portuguese,

Figure 14–22

who had indeed arrived on the Benin coast as early as 1485. We now know that this is incorrect. For one thing, the Portuguese did not have that skill in bronze casting. It is now almost certain that the art was brought to Benin from the still older culture of Ife, whose beautiful bronze heads only came to European knowledge in the 20th Century. It is, however, apparent that some of the designs appearing on the bronzes are of European origin (though Europe in turn learned them from the Near East). This is in contrast to the Bakuba patterns which seem to be purely African.

All seven strip patterns occur in Benin bronze work, but there are fewer plane patterns than in Bakuba art. Of the 17 possible plane patterns only 10 occur, and one or two of these are rather doubtful. Moreover, each type is represented by comparatively few different examples.

Figure 14–23 shows eight typical strip patterns, labelled according to their five types. As it happens, these all appear on the eight arms of a lamp, as recorded by H. Ling Roth in *Great Benin.*

(7)   (2)   (5)   (4)   (6)   (6)   (5)   (2)

Figure 14–23

Among the seven missing plane patterns in Benin art are the five that admit rotations of 120°. One of the two others is the design p4 which is also missing in Bakuba art. Figure 14–24 shows one of the more interesting plane designs which does occur. It is of type pgg, and probably has its origin in Portuguese armor.

pgg

Figure 14–24

4. *Adire Cloth*

The Yoruba *adire* cloth is made by starching a pattern onto white cloth, then dyeing the cloth blue before rinsing out the starch. In this way the starched portion remains as a white design against a blue background. Such cloths often consist of a number of squares each having its own pattern, and many of the 24 possible plane designs can be found there. Figure 14–25 shows six typical adire designs, labelled according to their type.

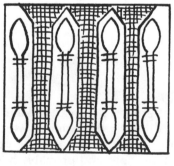

(7)

p1

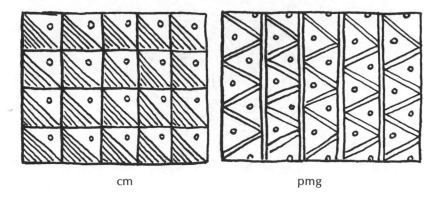

cm   pmg

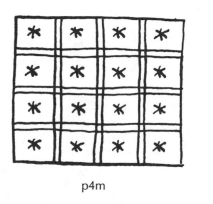   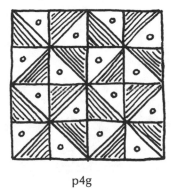

p4m   p4g

Figure 14–25

The starched pattern is often put onto the adire cloth with a stencil. When, after repeated usage, a stencil has breaks in it the starch makes a solid spot at the broken place. In that case a comb can be turned through the starch to make a decorative pattern like that of Figure 14–26 to cover the error. (Sometimes indeed this attractive design is put on intentionally.) It is interesting to note that this is exactly the construction which is used by Euclid in the very first Proposition of Book I of *The Elements* to construct an equilateral triangle having given side length—in this case the length of the comb. The vertices of the triangle are, of course, the centers of the two circles and one of their points of intersection.

Figure 14–26

SECTION 6
REGIONAL STUDY: SOUTHWEST NIGERIA

Recent excavations in the city of Ife, the religious, and earlier the political capital of the Yoruba peoples, have brought to light terracotta and bronze sculptures dating from the sixth to the twelfth centuries. The Africans who produced these works of art also developed an amazingly intricate system of numeration based upon twenty. Since early times they have engaged in commerce, using cowries as the most common currency. The number four is most significant in their cosmology. It appears in their division of time into four-day weeks as well as in the famous Ifa divination system.

Figure 15–1 Seventeenth century bronze plaque representing guards and attendants at a gate of the royal palace of Benin. Note the variety of pattern on the walls and on the guards' clothing and shields. These plaques "were kept like a card index up to the time of the Punitive Expedition (1897), and referred to when there was a dispute about courtly etiquette, as we know from an old chief who was a Palace attendant before the expedition" (Willett, page 105). British Museum.

CHAPTER 15
HISTORY OF THE YORUBA STATES AND BENIN

We are just beginning to discover the history of southern Nigeria, the most densely populated region of the whole continent of Africa. Since the peoples of this area left no written records, the key to their history must lie in archaeological excavations, works of art, oral traditions, language analysis, and reports of foreign visitors.

The recently developed technique of radiocarbon dating has brought infinitely greater accuracy to the field of archaeological research. By early 1967 thirty-one dates were known for all of Nigeria. It is estimated that five thousand dates are needed for the period 500 B.C.—A.D. 1500 to enable archaeologists to make an adequate reconstruction of the history of the country.

In the Nok area of northern Nigeria have been found stone implements whose age is estimated at more than 39,000 years. The iron-based Nok culture is believed to have existed in approximately 400 B.C.—A.D. 200; the terracotta art figures of this period are thought to have directly influenced the later art of the Yoruba people and of the adjoining city-state of Benin.

Concentrated efforts are being made to gather the oral traditions, but the difficulties are great. On the site of the former Oyo Empire of the Yoruba one now finds ruined settlements, some deserted during the Nupe invasion of the fifteenth and sixteenth centuries, some sacked by the Alafins (kings of Oyo) in the seventeenth and eighteenth centuries and many emptied by the nineteenth-century invasion of the Islamic Fulani from the north, as well as the Dahomean slave raids from the west. Lions occasionally greet the present visitors to these sites.

Essential to every African royal household was the professional historian, whose duty it was to memorize and recite the dynastic lists of the kingdom. His function was similar to that of the rhapsodists of the Homeric age. No doubt the facts were colored somewhat to please the current rulers; nevertheless, these lists constitute a fairly reliable record.

Toward the end of the nineteenth century the Rev. Samuel Johnson (Anla Ogun), a Yoruba Anglican minister, undertook to write down these oral traditions, so that the history of his fatherland might not be lost. Although educated Yoruba, schooled in the Western tradition, were familiar with the history of Greece and Rome, they knew nothing about their own land. Johnson spent over twenty years collecting his material and recorded it in *The History of the Yorubas*, which comprises the myths, history, and customs of his people.

Many traditions indicate that the Yoruba people came originally from the east. There are many similarities to the ancient Egyptian culture in religious observances, works of art, burial customs, and the institution of divine kingship; however, these cultural traits are shared by other African peoples. The ancient town of Ife, called Ile-Ife, was the spiritual center and cradle of the Yoruba people, and it was from Ife that the people of the neighboring state of Benin got their long line of kings. At this site have been found remarkable bronze, stone, iron, and terracotta works of art, dating back almost a thousand years.

One version of the creation legend begins with the Flood. Olorun, the supreme god, let his son Oduduwa down on a chain, carrying a handful of earth, a cockerel, and a palm nut. Oduduwa scattered the earth over the water, and the cockerel scratched it so that it became the land on which the palm tree grew, spreading its sixteen branches. The sacred number sixteen is believed to represent the sixteen crowned heads of Yorubaland.

According to another version of the creation myth, the Oba of Benin and the heads of the six Yoruba kingdoms were all descended from a common ancestor, Oduduwa. Oranyan, the youngest king, is supposed to have been the founder of the Oyo state.

By the year A.D. 1300, the Yoruba people had built numerous walled cities surrounded by farms. Trade was carried on with the peoples of the north; they exchanged cloth and kola nuts for products they needed. In turn the Yoruba states were exposed to the intellectual stimulation of the northern neighbors and the Islamic University in Timbuktu.

During the following centuries, the state of Oyo expanded until its dominion extended over a vast area, including even Dahomey to the west. Meanwhile the kingdom of Benin, to the southeast, had become independent of the Yoruba, and had grown into a mighty empire. When the Portuguese, the first Europeans to visit this part of Africa, entered Benin City in the late fifteenth century, they were truly astounded by the level of culture they encountered. Later European travelers described the thirty main streets of the metropolis, all straight and wide, laid out at right angles to one another, the longest extending four miles. "The people are as clean as the Dutch. Their houses shine like a looking glass," reported a Dutch visitor. The Oba's palace alone was reputed to be as large as a Dutch town. Brass-casting by the lost-wax method had been introduced from Ife, making possible the marvelous plaques on the walls of the palace, a history of Benin in art form.

With the Europeans came missionaries, gin, firearms, and the intensification of the slave trade. It was the wealth produced by the slave trade that helped to lay the basis for the expansion of the Benin and Yoruba empires. Every year the armies of the kings scoured neighboring territories in slave raids. In turn, their towns were attacked by other peoples with the same reprehensible motives.

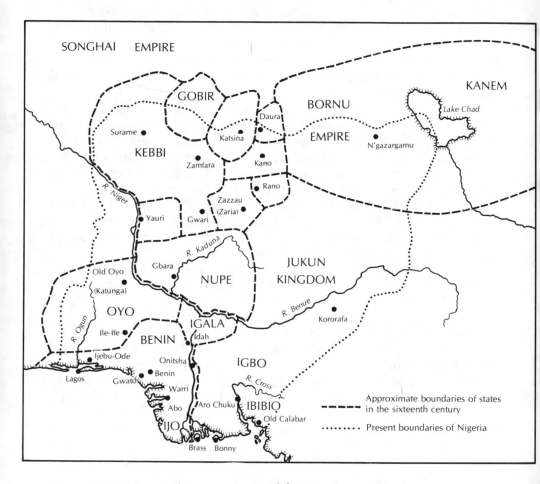

Figure 15–2   Sixteenth century states of the Nigerian region.

Slave raiding and wars with neighboring peoples alternated with periods of stability and peaceful trade—until the dissolution of the empires and the conquest by the British at the end of the nineteenth century. In 1897 a British delegation insisted upon seeing the Oba of Benin at a time when his religion forbade him to meet with strangers. Several of the British were killed in this incident. In retaliation the British sent a punitive expedition to burn the entire city. They removed 2500 beautiful Benin bronzes; these are now on exhibit in the museums of Europe and the United States.

The Yoruba city of Ibadan is the largest black city in Africa, a metroplis of more than 1,400,000 people, and the seat of the famous University of Ibadan. Although axes dating back to the Stone Age have been found in the area, Ibadan was only a small village built around a central market prior to the nineteenth century. During the time of the internal wars, it became a settlement for the army, a town free of the fear of invasion, a refuge from strife. Today three-quarters of the inhabitants are farmers who live in the city and work on the farms which surround it in a ring twenty miles wide. Actually, it is a city-village; farmers, craftsmen and traders live in large compounds in the town center.

Although Benin and the Yoruba states now form part of the republic of Nigeria, their language and culture remain very much alive. The Oba of Benin still reigns today. Modern Yoruba artists have adapted traditional Yoruba art forms to new subjects, producing works of great vigor and originality. As separate states within the Federal Republic, these enduring civilizations continue to flourish.

CHAPTER 16
SYSTEMS OF NUMERATION

The Yoruba System

"One of the most peculiar number scales in existence," wrote Conant of the Yoruba system of numeration. Indeed, one must be a mathematician to learn this complex system.

The Yoruba expresses forty-five, for example, as "five from ten from three twenties." In symbolic notation:

$$45 = (20 \times 3) - 10 - 5$$
$$106 = (20 \times 6) - 10 - 4$$
$$300 = 20 \times (20 - 5)$$
$$525 = (200 \times 3) - (20 \times 4) + 5$$

Quite a feat of arithmetic, involving addition, subtraction, and multiplication to express one number.

It is a vigesimal system, one based on twenty, of which we find many examples in western Africa. The unusual feature of the Yoruba system is that it relies upon subtraction to a very high degree. But to the Yoruba it seems perfectly natural, and he uses it with the ease with which we write IX (ten minus one) for nine in Roman numerals, or read the time as "twenty (minutes) to three."

A summary of the system is given by the Rev. Samuel Johnson in the introduction to his book, *The History of the Yorubas* (Preface, pages liv–lv).

> From one to ten, different terms are used, then for 20, 30, 200 and 400; the rest are multiples and compounds. Thus 11, 12, 13 and 14 are reckoned as ten plus one, plus two, plus three, and plus four; 15 to 20 are reckoned as 20 less five, less four, less three, less two, less one, and then 20.
>
> In the same way we continue 20 and one, to 20 and four, and then 30 less five (25), less four, and so on to 30, and so for all figures reckoned by tens.
>
> There is no doubt that the digits form the basis of enumeration to a large extent, if not entirely so. Five, ten, twenty, i.e., the digits of

one hand, of two, and the toes included, and their multiples form the different stages of enumeration.

Beginning from the first multiple of 20 we have *Ogoji*, a contraction of *ogun meji*, i.e., two twenties (40), *Ogota*, three twenties (60), *Ogorin*, four twenties (80), *Ogorun*, five twenties (100), and so on to ten twenties (200), when the new word *Igba* is used.

The intermediate numbers (30 having a distinct terminology), 50, 70, 90, 110, 130 to 190 are reckoned as: 60 less ten (50), 80 less ten (70), a hundred less ten (90), and so on up to 200.

The figures from 200 to 2000 are reckoned as multiples of 200 (400, however, which is 20 × 20, the square of all the digits, has a distinct terminology, *Irinwo* or *Erinwo*, i.e., the elephant of figures—meaning the highest coined word in calculation, the rest being multiples). . . .

By a system of contraction, elision, and euphonic assimilation, for which the Yoruba language is characteristic, the long term *Orun-din-ni* (*Egbeta* or *Egberin* and so on) is contracted to *Ede* or *Ode*, e.g. Edegbeta (500) . . . and so on. . . .

Summary. Thus we see that with numbers that go by tens, five is used as the intermediate figure—five less than the next higher stage. In those by 20, ten is used as the intermediate. In those by 200, 100 is used, and in those of 2000, 1000 is used.

Cardinal	Counting	Adjectival	Ordinal
1. okan	ookan = eni	kan	ekin:ni = ikin:ni = akoko
2. eji	eeji	meji	ekeji = ikeji
3. eta	eeta	meta	eketa = iketa
4. erin	eerin	merin	ekerin = ikerin
5. arun	aarun	marun	ekarun = ikarun
6. efa	eefa	mefa	ekefa = ikefa
7. eje	eeje	meje	ekeje = ikeje
8. ejo	eejo	mejo	ekejo = ikejo
9. esan	eesan	mesan	ekesan = ikesan
10. ewa	eewaa	mewaa	ekewaa = ikewaa

Table 5 Yoruba Numerals from One to Ten.

The figure that is made use of for calculating indefinite numbers is 20,000 (*Egbawa*), and in money calculation especially it is termed *Oke kan*, i.e., one bag (of cowries). Large numbers to an indefinite amount are so many "bags" or rather "bags" in so many places.

I have relied heavily for the analysis of Yoruba numerals upon Robert G. Armstrong's study, *Yoruba Numerals*. Armstrong states: "It is testimony to the Yoruba capacity for abstract reasoning that they could have developed and learned such a system."

Table 5 shows the words for the first ten numbers in the four principal applications. The double vowel with which each of the counting numbers begins is actually a contracted form of the word *owo*, meaning "cowrie" or "money." The original system, whose age is unknown, was expanded for the purpose of counting cowries.

The names of the numbers for the counting series, as well as their derivations, are given in Table 6. After 200 the system becomes quite irregular, and these irregularities are maintained on the higher levels.

The construction of the numerals from 35 through 54 is represented in Table 7. Of the twenty numerals, only five do not involve subtraction at all, and five numerals involve subtraction in two columns!

The idea of infinity is expressed in the Yoruba proverb: "There is nothing as numerous as the locusts; they are found at home and in the farm." Although some farms were as much as twenty miles from the towns in which the farmers lived, the locusts covered every bit of foliage on the route.

Dr. Akinpelu, of the University of Ife, gave me the following constructions for unit fractions, for doubling, and for powers of a base:

> *ebu* = fraction
> *Idaji* = one-half (divide into two)
> *Idameta* = one-third (divide into three)
> *Idamerin* = one-fourth
> *Idamarun* = one-fifth
> etc.

> *Ilopomeji* = two times, or doubling
> *Erin lona meji* = 4^2 (4 in 2 places)
> *Erin lona meta* = 4^3
> *Erin lona merin* = 4^4
> etc.

1 ookan

2 eeji

3 eeta

4 eerin

5 aarun

6 eefa

7 eeje

8 eejo

9 eesan

10 eewaa

11 ookan laa
(*laa* from *le ewa* = in
addition to ten)

12 eeji laa

13 eeta laa

14 eerin laa

15 eedogun
(from *arun din ogun* = five
reduces twenty)

16 eerin din logun
$(20 - 4)$

17 eeta din logun
$(20 - 3)$

18 eeji din logun
$(20 - 2)$

19 ookan din logun
$(20 - 1)$

20 ogun

21 ookan le logun
("one on twenty" =
20 + 1)

25 eedoogbon
$(30 - 5)$

30 ogbon

35 aarun din logoji
$[$five less than two
twenties $= (20 \times 2) - 5]$

40 ogoji
("twenty twos")

50 aadota
$[(20 \times 3) - 10]$

60 ogota
$(3 \times 20$, or more literally,
"twenty in three ways")

100 ogorun = orun
(20×5)

105 aarun din laadofa
$[(20 \times 6) - 10 - 5]$

200 igba

300 oodunrun =
oodun $[20 \times (20 - 5)]$

315 orin din nirinwo odin marun
$[400 - (20 \times 4) - 5]$

400 irinwo

2000 egbewa (200×10)

4000 egbaaji (2×2000)

20,000 egbaawaa (2000×10)

40,000 egbaawaa lonan meji
(ten 2000s in two ways)

1,000,000 egbeegberun (idiomatically
1000×1000)

Table 6   Yoruba Counting Series

	twenty	ten	unit
35	2	0	−5
36	2	0	−4
37	2	0	−3
38	2	0	−2
39	2	0	−1
40	2	0	0
41	2	0	1
42	2	0	2
43	2	0	3
44	2	0	4
45	3	−1	−5
46	3	−1	−4
47	3	−1	−3
48	3	−1	−2
49	3	−1	−1
50	3	−1	0
51	3	−1	1
52	3	−1	2
53	3	−1	3
54	3	−1	4

Table 7   Analysis of Yoruba Numerals 35–54.

Origin of the Subtractive System

Early investigators concluded that the evolution of this complex system for large numbers could be attributed to cowrie counting, the earliest occasion that required the Yoruba to count in such large denominations. In 1887 Mann described to a meeting of British anthropologists the procedure for cowrie-counting among the Yoruba. First the bag of 20,000 shells is emptied on the floor. The cowrie-counter kneels or sits beside the heap and rapidly draws four groups of five to make a small pile of twenty. Five twenties are combined into a pile of one hundred, and then two hundreds are swept together to form the important unit of two hundred. He attributes the subtractive principle in the numeration system to the practice of counting cowries by fives. Marianne Schmidl writes (page 94): "It is particularly evident in the case of the Yoruba that reckoning with cowrie shells was basic in the construction of the number system. With the single exception of the first decade, all numbers from five on are formed by subtraction from the next multiple of ten, which demonstrates that they were counted in fives and collected in stacks of ten and twenty units."

Although this method of cowrie-counting was widely used, few peoples besides the Yoruba and their immediate neighbors formalized the procedure in this unique subtractive system of reckoning.

Armstrong ventures the opinion that the pattern originated so that one could count the ten numbers with the fingers of one hand. If the multiples of ten are understood, and not shown by finger gestures, then one finger at a time can be extended to denote 21, 22, 23, and 24, respectively. When the fifth finger is extended, it is subtracted from 30; and when it is retracted, the remaining four fingers are deducted from 30 to give 26, etc.

Yoruba Record-Keeping and Computation

Thus far I have not been able to determine what computational methods were used in past centuries by Yoruba merchants in commercial transactions involving many thousands of cowrie shells. Nor have I yet found whether they kept written records of debts, or whether trade was strictly on a cash basis.

Dr. Ajayi, of the University of Ibadan, writes that computations are usually performed mentally. Records of numerical transactions are "sometimes marks on walls, beans or sticks in a jar, etc. to represent large units of what is being counted; e.g., a hundred or thousand cowries or heads of cattle, etc. owed, or years of a monarch's reign. In some kingdoms, censuses of the number of families in each village or district are kept in the same manner."

Yoruba Children Learn Their Numbers

Chief Ayorinde, the authority on Yoruba folklore, commented upon the teaching of numerical concepts:

The knowledge of numerals among the Yoruba People is as old as life itself. It is usually taught or imparted in different forms from childhood or as soon as a child can recognize an object. Pebbles, beans or stones are sometimes used in the teaching of numbers through demonstration. This teaching is also continued through traditional games as in Ayo, otherwise known as the "Warri Game," and it often sharpens one's mental capacity for calculations.

Children also learn the art of buying and selling while in company of their mothers to the market. Apart from helping with the supervised sales of the parent's articles, the children are sometimes given a few wares of their own so as to give them confidence and as a test of their understanding of money values. The teaching of mathematics could also be formal and informal, indoors and outdoors, and takes many forms.

Today in this country, commercially produced materials are designed by educators to give the American child concrete experience with numbers—the kinds of experiences that are a normal aspect of Yoruba upbringing.

Updating the Yoruba Numeration System

During the colonial period, the languages of the dominating European powers were used in African schools, except possibly in the primary grades. Since they gained independence, however, the African peoples have been eager to revive interest in their own languages and culture; hence there is a new desire to expand the native vocabulary to include concepts hitherto unexpressed. Although English is still used today in Nigerian institutions of higher learning as a common means of communication for students of varied linguistic backgrounds, the Yoruba language is being expanded to accommodate new ideas.

Dr. Armstrong, the Director of the Institute of African Studies at the University of Ibadan, has developed a decimal number system which uses Yoruba words throughout and eliminates all the irregularities of the traditional system. Thus modern arithmetic would become readily accessible to average people and to youngsters in the elementary grades, without requiring that they learn a language other than their own. Dr. Armstrong's reforms include terms for all the arithmetic operations, as well as fractions, decimal fractions, and percents. Dr. Armstrong has confidence that these reforms will encourage people to take a great interest in mathematics, as they realize they are citizens of an increasingly mathematical world. He concludes: "... a nation that could develop and transmit the complexities of their [numeration] system has obviously a great deal of mathematical talent and interest" (page 22).

The Numeration System of Benin

The construction of the number words in the Edo language, spoken by the people of Benin, is similar to that of Yoruba, with the exception that in Edo the words for numbers having five in the units place are formed by addition, rather than by subtraction. Take, for example, the words for 45 and 46.

$$45 = (60 - 10) - 5 \qquad \text{in Yoruba}$$
$$45 = 40 + 5 \qquad \text{in Edo}$$
$$46 = (60 - 10) - 4 \qquad \text{in both Yoruba and Edo}$$

In Edo the first five numbers following a multiple of ten are formed by addition to that multiple of ten, while the next four are based on subtraction from the next higher multiple of ten.

In his paper, "How the Binis Count and Measure," Peter Idehen gives this interesting interpretation of the word for fifteen: "*Ekesugie* means either 'mid-way between *Igbe* (10) and *Ugie* (20)' or 'five out of twenty.'" For twenty-five he gives two forms:

25   *Isen-yan Obgen* (5 and 20) or *Ekesogban* (midway between 20 and 30)

I have at hand four sources for the Edo number words, all differing from one another. Since the most recent is that of David A. Munro (1967), I am using this book as the basis for the numeration table, with some modifications suggested by Mr. Idehen's paper (Table 8).

On the subject of counting beyond one hundred, Idehen writes that in the past most people either had no occasion to use such large numbers, or else they resorted to English. He urges that people learn to count in Bini up to the millions, so that they can translate into their own familiar language, for instance, a radio broadcast stating that the government plans to spend two million pounds on a project.

For the higher numbers he gives:

1000	*aria isen* (5 × 200)
2000	*aria isen eva* (2 × 1000)
10,000	*aria isen igbe* (10 × 1000)
1,000,000	*aria isen aria isen* (1000 × 1000)
1,000,000,000	*aria isen aria isen aria isen* (1000 × 1000 × 1000)

1	owo or okpa	16	ener-ovb-ugie (4 out of 20)
2	eva	17	eha‿iro vb-ugie (3 out of 20)
3	eha	18	eva‿iro vb-ugie (2 out of 20)
4	ene	19	okpa‿iro vb-ugie (1 out of 20)
5	ise	20	ugie
6	eha	30	ogba
7	ihiro	40	iy-eva (20 × 2)
8	erere	50	(e)k-igbe s-iy-eha $[(20 \times 3) - 10]$
9	ihiri	60	iy-eha
10	igbe	70	(e)k-igbe s-iy-ena $[(20 \times 4) - 10]$
11	oworo	100	iy-ise (20 × 5)
12	iweva (10 + 2)	190	igbe‿iro vb-uri (10 out of 200)
13	iwera	200	uri or aria okpa
14	iwene	300	uri okpa vb-iy-ise or iyisen yan uri (100 + 200)
15	ekesugie	400	uri eva or aria eva (200 in 2 places)

Table 8   Edo Number Words

The city of Ife is the spiritual and cultural center of the Yoruba people. Here archaeologists have found beautiful cast brass and terracotta heads, masks and figure groups, and figures in stone, dating back to the thirteenth century. Here still stands the "staff of Oranyan," the stone obelisk marking the burial spot of the legendary founder of the Yoruba kingdom of Oyo. Here one also finds the ancient shadow clocks, stone monoliths that served to tell the time of day and fix the dates for festivals. And here the Oba of Benin was given token burial, even though the kingdom of Benin had achieved independence from Yoruba domination. The influence of the religious beliefs of Ife spread to Benin, to the Igbo, to the Nupe, and elsewhere.

Significance of Four

Among these peoples certain numbers are held sacred, and the number four occupies the principal position among them. Thus we find that a four-day week is traditional among the people of southern Nigeria. Among the Yoruba each day is dedicated to one of the four major deities or its local counterpart: Shango (Sango, Jakuta), an early mythical king, later worshipped as the god of thunder, is represented by the double-axe symbol of the lightning bolt; Obatala (Orisala, Orisha-nla), the creative god, coworker with the supreme god Olorun, the god of purity, is protector of the town gates and patron of the physically deformed; Orunmila is the oracular deity of the Ifa divining system; and Oduduwa (Ogun), the legendary founder of Ife, is god of war and patron of iron-workers.

"The world has four corners" is a widespread Yoruba expression; similarly, the approach to the earth from the outer world is through four gates. This image is carried out in the four gates of the walls of each town. One of the Yoruba deities is Olori Merin, the four-headed being, whose image was set up in a prominent place in the town so that each head faced one of the four points: east, north, west and south. The four major deities, whose names were given to the days of the week, were linked with these points; Shango with the east, Obatala with the north, Orunmila with the west, and Oduduwa with the south. Each of the four directional quadrants was divided again into four parts, each being under the patronage of one of the sixteen deities who were responsible for the formation of the earth.

The days of the week in Benin are also associated with the four quarters of the earth. Eken, the day of rest, relates to the east, Orie to the west, Aho to the south, and Okuo, the day on which the Oba performs his ceremonies, is linked with the north. In the Igbo language the names are similar: Eke, Oyo, Afo and Nkwo.

Figure 17–1   Yoruba beaded boots, Nigeria. According to tradition Oduduwa, the first Yoruba king, gave a beaded crown and other items of apparel to each of his sixteen sons, who subsequently became the rulers of the various Yoruba states. Among the items were beaded boots with four stuffed beaded birds running up the front of each. The boots in the photograph belonged to a chief whose claim to them was rejected, thus forcing him to forfeit them. British Museum.

Significance of Four in Divination

Of great import are the powers of four: 16, or 4^2, and 256, or 4^4. These numbers are central to the art of divination known as Ifa.

The Yoruba people are renowned for their Ifa divining system, the means of consulting the oracular deity Orunmila. The advice of the god is solicited prior to every important step involving a decision. This is done by means of sixteen nuts of the palm tree, whose importance is unique in the life of the people. A seventeenth nut might lie nearby in a ring of cowrie shells, called "money of Ifa." The diviner first shakes the sixteen nuts in his two hands; this process is called "beating the palm nuts." He then places them all in his left hand, and tries to pick up as many as possible with his right hand. If exactly one nut remains in the left hand, he traces two strokes on the powdered surface of the beautifully decorated Ifa tray; if two nuts remain, he makes one stroke on the Ifa try. A record is made only when exactly one or two nuts remain in the left hand.

Figure 17–2   Bowl for the palmnuts used in Ifa divination. Dahomey. British Museum.

This whole process is repeated until eight single or double strokes have been recorded in two columns on the tray. One arrangement is shown in Figure 17–3.

Figure 17–3   Ifa divination system of the Yoruba of Nigeria and Dahomey. The arrangement of strokes in two columns on the powdered Ifa tray refers the the diviner to the appropriate odu, or verse of the oral literature, in which the client can find the answer to his problem.

Each column has a distinctive name and significance, and since there may be either one or two strokes in each position, there are 2^4, or 16 different arrangements in each column. The names vary from region to region. As given by Bascom, four arrangements are:

Ogbe	Oyeku	Iwori	Edi

The right column is considered male and is more powerful than the left, or female, column. Each of these double-column arrangements is called a "road of Ifa" or an *Odu*. Altogether there are $2^4 \times 2^4$, or 256 different *odus*, which are ranked from 1 to 256 in the order of their importance. With each *odu* there is associated a number of verses, each verse being related to a problem in real life, but couched in the oblique language so characteristic of Yoruba. As the diviner recites the verses associated with the *odu*, the client listens for the words appropriate to his own situation. The priests are also renowned as skilled physicians.

The Rev. Johnson suggests that the Ifa system resembles that of the oracle at Delphi. The following verse indicates a similarity to present-day psychoanalysis; it gives the suppliant the confidence necessary to carry out the desired undertaking—and exacts a price (Bascom, page 143).

Orunmila says each should take his own row;[1] I say each should take his own row; he says that Twenty Cowries takes his own row but cannot finish it.

Orunmila says each should take his own row; I say each should take his own row; he says that Thirty Cowries takes his own row but cannot finish it.

Orunmila says each should take his own row; I say each should take his own row; he says that Forty Cowries takes his own row but cannot finish it.

I say, "Well then, my father Agbonnire,[2] who can complete his row?" He says Fifty Cowries alone can complete his row, because we cannot count money and forget Fifty Cowries.

Ifa says he will not allow the person for whom this figure was cast to be forgotten. This person wants to do something; he will "complete his row" in the thing he wants to do.

A simpler system of divination, used for minor decisions, calls upon Opele, a lesser deity. For this purpose the oracle casts a divining chain on which eight half-pods or coins are strung at regular intervals, but with a greater space between the fourth and fifth symbol. The diviner picks up the chain in the middle and lets it fall upon a flat surface in a two-column arrangement. The appropriate odu is determined by the sequence of convex and concave surfaces of the half-pods, or heads and tails on the coins, and the interpretation is exactly as with the palm nuts.

Some authors, including Ojo, claim that there are 16^3, or 4096 odus (roads of Ifa), while others state that there are as many as 16^4, or 65,536. Bascom and Idowu maintain that the number is precisely 256.

For centuries the Yoruba rulers have maintained a hierarchy of Ifa diviners. At one time these oracles held a powerful position in the government of the state, even to the responsibility for the deposition of the reigning Alafin. Europeans reported witnessing the practice of Ifa at least as far back as the year 1705. The cult has spread far and wide—to Dahomey,

---

[1] The row refers to a row in the fields to be hoed or weeded. Here, and elsewhere in these verses, it is often used in a broader sense to mean any undertaking, so that "completing one's row" means being successful in a given venture.
[2] Agbonnire is a shortened form of Agbonniregun, another name for Orunmila or Ifa.

Togo and parts of Ghana. Ifa geomantic divination has been preserved among the Yoruba descendents in Cuba and Brazil, where it has merged with Catholicism, and a Yoruba temple was recently opened on Seventh Avenue in New York City.

Other Significant Numbers

The number sixteen, obtained by raising four to the second power, has significance outside of its central position in divination. For example, sixteen occurs in one of the creation legends, that of the palm tree with its sixteen branches representing the sixteen kings of the Yoruba.

Other important numbers are 40 and 200, as well as their successors 41 and 201.

The state funeral of Lt. Col. Adekunle Fajuyi in January, 1967 was marked by the chanting of the Ekiti traditional dirge. At one point the singer cried:

> I call you, won't you please answer?
> I call you five times, six times!
> I call you seven times, eight times!
> I call you sixteen times. . . .

Later the cry was repeated in a different form:

> Your father calls you five times, six times!
> He calls you seven times, eight times!
> He calls you sixteen times
> Where the Olubije-mushrooms grow all over the road!
> He calls you without stopping!
> Your mother calls you too.

Reference was made to the sacred number 200:

> "The one who enlightened 200 persons is your father."

Included in the gifts which a young man presents to his betrothed are forty kola nuts. Ritual requires that they be split and divided among the guests, indicating that they have witnessed the betrothal. The symbolic image associated with Ori, the household deity, is a crown of forty-one cowries.

The sacred numbers of the Bini (people of Benin), according to Peter Idehen, are "20, 2, 4, 7, 14, 40, 200 and 201. The Binis usually give kola-nuts to strangers in twos, or fours. If the number of people is great, they give

7, 14 or 20. If the number is greater, as for example, a whole village, they give 200 or 201. When they offer sacrifices to their gods, they follow the same order. Only very rich people or the Oba can offer 200 or 201 kolanuts or sacrificial victims."

The Yoruba recognize many subsidiary gods, or orishas (orisas), each associated with a specific function or event. The Rev. Johnson states 401 as the supreme number in the hierarchy; other authorities claim 201, 401, or 1600.

The worship of Orisha-nla, the deity of purity, has an important place in Yoruba life. The god's responses are thought to be interpreted by casting the four-segmented kola nut. It can fall in one of five different positions, each carrying a certain symbolism. Sacrifice to this Orisha includes sixteen snails, sixteen roasted rats, sixteen dried fish, and sixteen kola nuts. In another phase of the festivities the priests strip 201 leaves from the branches of certain trees.

The number seven occurs in connection with the seven-day harvest festival, and with the Egugun celebration, which is dedicated to departed ancestors and observed with great homecoming festivities. One version of the creation myth refers to the seven grandchildren of Oduduwa, who subsequently became the rulers of the Yoruba and Benin peoples. Initiation into the mysteries of the worship of Shango, the god of thunder, required the payment of seven head of cowries (14,000 shells), according to Johnson.

Fourteen Hundred

"Which is the very long coffin that can accomodate 1400 corpses?" The answer to this old Yoruba riddle is "the path," depicted as a long coffin containing many bodies arranged end-to-end in single file. This image represents the hundreds of farmers who file along the paths on their way to and from their farms during the morning and evening "rush" hours! The number 1400 is composed of the factors 7 and 200, both significant numbers.

"Fourteen Hundred Cowries" is the tale of a cricket who set out to pay his future father-in-law the stipulated bride-wealth of 1400 cowrie shells. The cricket first had to borrow the money for the payment. On the way he encountered a wine seller, and before he knew what had happened, the cricket was drunk. Then followed a series of adventures, until at last the King's son was involved. The last few verses sum up the story. As the King gave his son the 1400 cowries, he said:

> "O son, as you say, this trouble is great.
> From my treasury funds we surely must pay.
> Many of my subjects are heavily in debt.

>First 1400 to you for the drummer,
>Then the drummer pays the woman this money.
>The woman passes on the cowries to the hawk . . ."

Hawk, hen, women, stump, hunter, roan, cotton tree, and finally:

>"If the cricket isn't drunk, he pays the wine seller
>The 1400 cowries from my treasury,
>The gods will be angry if we do not pay up."

By the time the 1400 cowries had been given to the cricket, he had recovered from his drunkenness. He paid for his wine, and then set out again to borrow 1400 cowries for the bridewealth.

Oranyan's Staff

A fascinating subject is the origin and meaning of the symbols carved into the obelisk to the founder of Oyo, the "staff of Oranyan," still standing in Ile-Ife. The Rev. Johnson writes that the symbols are רי, the Phoenician letters Resh Yod (read from right to left), representing the name "Oran-yan." Precisely these letters were used in classical Hebrew, which was derived from Phoenician, as symbols for the numerals 200 and 10, respectively. The sacred number 200 again!

If we go further back to the Egyptian hieroglyphs, we discover that the letter Resh can be traced to the image of a head, while Yod represents a hand. Are these letters possibly linked with Ra, the sun-god of the Egyptians, and with Yahweh, the forbidden name of the Hebrew god? Recent investigations dash cold water on any such speculative interpretation. The eminent authorities, William Fagg and Frank Willett, in their 1960 article "Ancient Ife," state that there is "no justification at all for the interpretation of the relief carving on the Opa Oranmiyan in terms of Hebrew or Egyptian characters, nor for any explanations offered to account for the iron nails." They claim that the sketch of Oranyan's staff in Johnson's The History of the Yorubas bears little resemblance to the original, and that the number of iron nails driven into the granite monolith does not agree with that given in Johnson's description. It may take many more turns of the archaeologist's shovel to discover the true genesis and significance of the obelisk.

The day begins at dawn and ends at twilight. The night period separates one day from the next. In the center of every Yoruba compound are ever-green shrubs and forked sticks holding pots of sacred potions. These objects serve as sundials; from the length of the shadows at various times of the day the Yoruba is able to estimate the time.

In Ile-Ife there stands an ancient sun-clock, a triangular monolith around which a circle has been drawn. This circle is divided into sixteen parts, and the position of the shadow in relation to the circle indicated in earlier days the time for the observance of certain festivals.

When the sun is hidden by dark clouds, the farmer estimates the time by the amount of work done, by the cries of certain birds, and the feeding behavior of the animals. Peter Idehen writes: "Birds like Erhimohi, Awe, and Okpa (cock) help to tell the time. The bird called Erhimohi tells time by its cries at night. It cries at six o'clock in the evening, at twelve o'clock midnight, and at four and six o'clock in the morning. In most cases this bird cries at the correct time even when checked with the clock. The bird called Awe cries every four hours, and like the Erhimohi, can be fairly accurate as a time keeper. The cock cries as soon as the morning begins. In many villages and homes the cock will be killed almost immediately if it crows from seven to two o'clock at night, for people suspect some dangers. The use of the shadow and these clock-birds serves the people as clocks serve Europeans."

Although the European calendar with its seven-day week has been adopted officially in Nigeria, some aspects of life in southern Nigeria continue on the traditional four-day basis—most notably, the markets. In the past, the most popular god in a locality marked the beginning of the Yoruba week in that particular area. Later it was the occurrence of the market in that locality which governed the ordering of the four days: market day, market's second day, market's third day, tomorrow is market day, and finally another market day. An Igbo speaks of the "Eke market in the village of Okperi"—the market held every week on the day called "Eke."

Time-reckoning is inclusive: when one speaks of a certain period of time, one includes both the first and the last unit in the interval. Markets are said to take place every five days or every nine days; in other words, once a week or once in two weeks, based on the four-day week. The Yoruba expression for "two years ago" is *Idunta*, derived from *Odum meta*, literally "three years." In *Arrow of God*, an Igbo Christian says: "This is the God about whom we preach every eighth day," referring to the seven-day week. This practice of counting inclusively led some European investigators to the conclusion that the Yoruba actually had a five-day week, with an unnamed day of rest.

The month is reckoned as the interval between the two consecutive appearances of the new moon. Accordingly it has twenty-eight days in the Yoruba calendar. This number is actually closer to the sidereal month, calculated with reference to a particular star. A Yoruba friend recalls that his father made marks on the wall to keep count of the passing of the days in each month.

On the concept of the year, Ojo writes (pages 205–206):

The ancient Yoruba idea of a year probably passed through three important stages in its evolution: the farm cycle, the season cycle, and the corresponding lunation cycles. Like man in any environment who must eat before he can philosophise, the Yoruba counted time in connection with his food production. In those early days of settled farming his normal calendar was derived from his farming activities which were in themselves rhythmic . . . Later the year was viewed as a succession of the predominant weather obtaining during the two main seasons, wet and dry. . . . The connection between the lunar cycles, the season cycle and the farm cycle was not overlooked, and it led to the Yoruba estimate, based on practical observation, that thirteen lunar months make one year.

The calendar of Benin is similar to that of the Yoruba. Mr. Idehen tells of the role of the King's astronomers:

The Bini have a certain society supervised by the Oba and whose duty is similar to that of the Astronomers. The society is called Iwo-Uki (The Rising Moon). Their duty is to watch the moon, the sun and the stars and tell the people the time for the various festivities, time for planting, weeding, harvesting and so forth. . . .

The Society observes certain birds and insects to help them to determine the seasonal changes. The bird like the wagtail or agret is only found during the dry season; so that as soon as they are seen people know that dry season is approaching. The presence of insects called the crickets and other insects in the forest are also indicative of seasonal changes. They also watch the stars to determine seasonal changes. The most important star is the Orion's Belt (Agho). When this star disappears from the sky the people know that it is time to plant their yams. Though these devices may seem primitive to the "civilized" people, yet the people had used them with some degree of accuracy. The devices are natural and are the result of many years of careful observation.

Longer periods of time are reckoned in relation to important events, and these in turn are coordinated with the land rotation system in each locality. The actual number of years in the cycle depends upon the number of plots available, and in the not too distant past it varied from five to twenty years. By employing this method of calculation an elder could recall the ages of his children, the date of a smallpox epidemic, or an invasion by locusts.

Dr. Ajayi sums up the Yoruba traditional time-sequence:

A year is usually reckoned in terms of annual festivals and seasons, and in relation to the reign of particular rulers. Within the year, there are divisions into lunar months, dry and rainy seasons; times of festivals to different gods which form an annual cycle, and each of which is determined also in relation to the seasons, agricultural time table of planting, clearing and harvesting, as well as market days. Market days are regular, every fifth, ninth or seventeenth day, and thus provide the basis for fixing precise dates. Besides regnal years, there are other bases like the formation of Age-grade associations, e.g., every third year; each association bears a different name related to the reigning monarch.

CHAPTER 19

MARKETS AND CURRENCY

The people of southern Nigeria have been known for centuries past for their proficiency in commerce. This reputation extends to both men and women. Markets were—and still are—held at regular intervals, usually every four or eight days to coincide with the traditional four-day week. The location and the products of the local markets rotate. Today the cloth market in Ibadan, held every sixteen days, attracts customers from points as distant as Ghana.

Since the market women combine the roles of housewife, food producer, trader and customer, the periodicity of the market is a necessity. Often the women are in charge of trade, ranging from village or pavement-level buying and selling to the big business of the Lagos markets. Yoruba women have a reputation for knowing where to buy cheap and sell dear. They go directly to the farms several miles from town, or wait on the farm paths to purchase produce on its way to the local market.

A recent article in TIME magazine extolled the perspicacity of the West African market women, who control much of the transport and trades in textiles, food and hardware in Ghana and Nigeria. Bankers in Lagos tell of one woman who cannot write her own name, but who can get a letter of credit for 200,000 pounds ($560,000) whenever she needs one. To quote one African journalist: "They can't read or write, but they can damn well count."

Cowrie Shells in Southern Nigeria

Although goods were often exchanged by barter arrangements, the merchants of southern Nigeria had adopted currency long before the arrival of the Europeans. As the number words indicate, the cowrie was the basic unit of currency. Large quantities of them have been excavated at ancient Ife.

When the Portuguese paid their first visit to Benin in the fifteenth century, they found cowrie shells in common use. In 1588 James Welsh, an English trader, bought two gallons of palm wine in Benin for twenty shells. A year later two London merchants named Bird and Newton reported that "pretty white shells" were used as Europeans used gold and silver. They were able to purchase two gallons of honey and a honey-comb for a sum of one hundred cowries.

About the year 1700 a Dutch trader told of the royal monopoly of the Oba of Benin. No specific duties were imposed on imports or exports, but there was an annual tax in cowries for the privilege of trading. Each territorial governor was required to raise a specified number of bags of boesies (cowries) for the Oba's use, while officials of lower rank delivered produce for the royal household. There were no poor people in the land; the wealthy were obligated to see that none went hungry in this feudal empire.

How were cowrie shells handled? How does their value compare with that of European currencies? I have relied most heavily on Marion Johnson's fine analysis, "The Cowrie Currencies of West Africa" and on A. H. M. Kirk-Greene's excellent article, "The Major Currencies in Nigerian History."

In the northern parts of the region, the cowrie shells were counted out in groups of five, while along the coast they were pierced and threaded, generally in strings of forty. It is not known whether cowries were strung at the time the Portuguese first began to trade. We know they were counted in units of forty, named *galinhas* (Portuguese for "hens"). Later this name was applied to a bunch of five strings, or two hundred shells, no doubt to correspond to the devaluation of the cowrie in terms of its purchasing power. As the commercial language changed from Portuguese to English, later equivalents were given in English.

In areas where cowries were not strung, their use depended upon a rapid method of grouping them in successively higher units, as described in the chapter on Yoruba numeration. Let us remember that in Nigeria trade was carried on by both men and women, and that cowries had to be counted up into the high denominations. Furthermore, they were counted out by both the buyer and the seller! Certainly this should have dispelled the myth that Africans could barely count to ten.

The early Yoruba system was based on a combination of vigesimal and decimal counting: 20, 200, 2000, and 20,000 (*oke kan* = one bag). The system at Lagos was later adopted further north:

$$40 = 1 \text{ string}$$
$$2000 = 1 \text{ head} = 50 \text{ strings}$$
$$20,000 = 1 \text{ bag} = 10 \text{ heads.}$$

However, the 200 unit was recognized; a discount of two cowries was allowed on every five strings.

The Yoruba, Osifekunde, who was sold into slavery in Brazil in the early nineteenth century had spent his youth in the Ijebu area, northeast of Lagos. His recollections of life in his homeland included remembrance of the currency values. He gave the following cowrie equivalents:

> ogoji   = string = 40 cowries
> ogwao   = bunch (of five strings) = 200 cowries
> egwegwa = head (of ten bunches) = 2000 cowries
> oke   = bag (of ten heads) = 20,000 cowries.

An average price for a slave at that time was two bags.

All large trading centers in the western Sudan area employed cowrie counters. Imagine having to count daily up to 300,000 of these small curved shells, particularly in the inland regions, where they were not strung.

In the 1860s the cowrie table and the British equivalents read (with variations depending upon time and place):

> 40 cowries   = one string = ¾ − 1 penny
> 5 strings   = one bunch = 3 − 6 pence
> 10 bunches = one head = 1¾ − 2 shillings
> 10 heads   = one bag = 14 − 18 shillings.

By the end of the century, one thousand cowries were worth three pence in silver, but a copper penny could be exchanged for only 300 cowries.

During the early nineteenth century, the larger Zanzibar cowries were introduced. European merchants found they could scoop them up by the ton and sell them at a handsome profit. With the added weight of the shell and the depreciation in value owing to the introduction of European currency, by the end of the century it was hardly worthwhile to transport them. On the lower Niger River the cowrie became merely a measure of value—prices were quoted in so many cowrie units. It had become impractical to use them as a medium of exchange except for small purchases in the local markets. The fluctuations in the value of the cowries and the need to convert to British units must have taxed the skills of the shrewdest of merchants. As an example, in 1902 the rate was four thousand to a shilling in the Yoruba region, while at Sokoto, in northern Nigeria, a shilling would fetch only 1200 shells. Imports of cowries were banned by Proclamation in 1904 and coinage was introduced by the government. However, since the smallest unit was a threepenny piece, cowries continued to be used in local trade until the introduction of the anini, worth one-tenth of a penny.

The subtractive principle is used in the following constructions in Yoruba:

two (shillings) and threepence = *(m)eji le toro*
two (shillings) and ninepence = *(m)eta din toro*
(three less threepence)

The cowrie units remain useful as a means of expressing large numbers. A farmer might say his farm has "three-pence" (1000) yam heaps. In earlier times it was recorded that the King of Dahomey was beaten by the Yoruba with a loss of "two heads, twenty strings, and twenty," or 4820 soldiers.

Cowries in the Twentieth Century

By no means did cowrie shells disappear after the British government introduced coinage. Cowries are an integral part of daily life in many regions of Africa, for use as decoration, religious symbols, and special-purpose currency. Even as ordinary money they were still used in recent times. In the 1920s the Igbo people kept them in circulation, particularly in the inland areas. In Yoruba and Nupe territory they reappeared during the severe depression of the 1930s, when even the *anini* (one-tenth of a penny) was too large a unit of exchange. As late as 1942 payments in some parts of Nigeria were expressed in cowries, rather than in coinage.

Although the use of cowries as ordinary currency has been discouraged or outlawed, these small shells have a function as special-purpose money—as bridewealth and for various ceremonial payments.

At the time Samuel Johnson wrote his book, wealthy families required more than ten heads of cowries (over 20,000), as compared with earlier times when a token payment of one head was considered ample. Today the brideprice is generally tendered in cash. In 1969 maximum brideprice payments were set by the government in some parts of Nigeria, since the matter was getting out of hand—the higher the girl's education, the greater the brideprice!

Among the Yoruba ceremonial occasions which require cowrie payments are funerals, initiation into secret societies, and certain fines. As decoration, cowries are seen everywhere—on clothing, drums, divining chains, headdresses, ritual masks, and furniture.

Other Forms of Currency

We shall mention briefly the various articles which have been used as currency in European trade with the territories that today make up Nigeria. In

1510 one could buy a slave for eight or ten copper *manillas* (Portuguese for "bracelet"). There were five different patterns of manillas in Nigeria, each accepted only in certain areas, and in use into the present century. By the end of the nineteenth century, five standard manillas bought one bottle of gin.

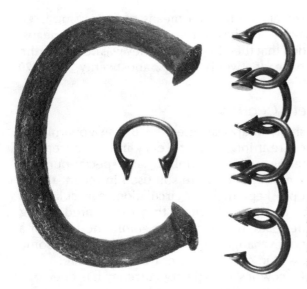

Figure 19–1 Manillas, the most celebrated currency in the Niger River delta, varied in weight from three ounces to six pounds. British Museum.

Copper, iron and brass rods were in great demand. They were manufactured into ornaments, works of art, weapons and tools, and special-purpose currency. The Rev. Johnson complains that "soon doubts will begin to be expressed as to whether Yorubas ever knew the art of smelting iron from the ores!" The iron bar gave rise to the terminology "a bar of tobacco" or "a bar of rum." In southern Nigeria the *Ogoja* penny, a Y-shaped iron bar, was worth about a half-penny. Basden speaks of a unique currency in one part of the Igbo region, tiny pieces of iron with arrow-shaped heads, used in olden times for the acquisition of slaves. "How many thousands had to be counted when making such a purchase baffles one's imagination." The higher currency used in this area was brass rods. Beads were widely reported as currency, and valued for their decorative attributes. The Rev. Johnson reports that when one went on a long journey, one took beads instead of cowries.

Of all the currencies in use in the nineteenth century, only guns and alcohol achieved any stability of exchange rates, according to the African historian K. O. Dike. Bottles of gin passed from hand to hand for years without having been opened, and might represent the entire wealth of a chief. Basden reported seeing huge collections of empty gin bottles—a record of past transactions! The British had encouraged a taste for liquor among the people on the African coast, since it constituted a cheap and convenient medium of exchange.

At all times trade was carried on by barter. To mention just one such exchange, involving human "cargo," a British ship in 1676 bought one hundred slaves, all branded with the mark of the Duke of York, for various lengths of cotton cloth, five muskets, twenty-one iron bars, seventy-two knives, and a half barrel of gunpowder.

SECTION 7
REGIONAL STUDY: EAST AFRICA

The key to East African culture is cattle. The inhabitants of the region stem from various parts of Africa and speak a diversity of languages, but for most of these people, cattle are their most valuable possessions. Commodity values are expressed in terms of livestock, and ultimately, of the prized cattle. They are used for bridewealth, to make reparation for crimes and insults, as religious offerings. The division of the day in many societies is based on the sequence of activities with regard to cattle. The taboo on counting the animals in the herd is circumvented in many cases by resorting to finger gestures.

Figure 20–1   Kilwa coins minted during the reign of Sultan Ali ibn Al-Hasan, 885–887 A.H. (1480–82 A.D.) The Arabic inscriptions are: Side A. "Ali son of Al-Hasan, May he be happy." Side B. "Trusts in the Master of bounties." National Museum of Tanzania.

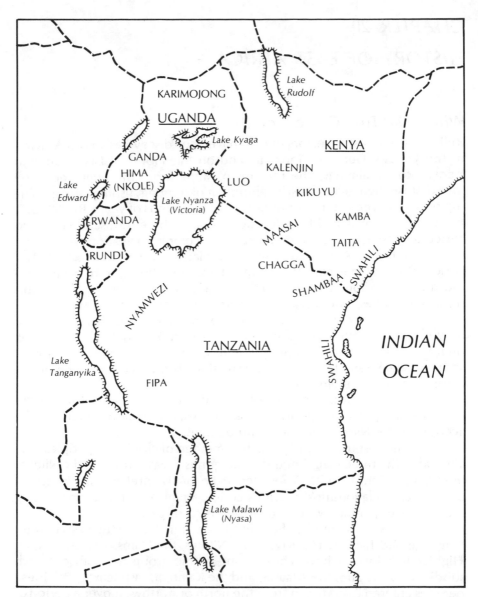

Figure 20–2   East Africa: Countries and Peoples.

CHAPTER 20

HISTORY OF EAST AFRICA

Where Did They Come From?

Relics of the oldest ancestors of man have been discovered in East Africa—in the Olduvai Gorge in Tanzania and on the shores of Lake Rudolf in Kenya. Rock paintings, the remains of stone tools, the engraved bone found at Ishango, tell of the development of hunting, fishing, and gathering societies over a period of hundreds of thousands of years. By 5000 B.C. herding of animals had been introduced, and three thousand years ago agricultural societies developed, with their complex irrigation systems and terraced hillsides. About a thousand years later iron-working came to East Africa. The source is not clear; it may have been brought from Meroë in the north by Nilotic peoples, or from the south and west by the Bantu-speaking peoples, who began to penetrate East Africa at that time.

The history of East Africa is one of migrations and changes. People moved to find better lands, to escape the disruptions of war, or to seek higher status in a new community. For the most part, movements were over short distances, and the existing population absorbed the newcomers, or were absorbed by them.

The present inhabitants of East Africa are of many ethnic backgrounds and speak a great variety of languages. This section contains a brief account of the Nilotic and the Bantu peoples.

The most useful and objective method of classifying peoples is by language. The three main linguistic categories in East Africa are the Nilotic, the Bantu and the Cushitic. They seem to be totally unrelated. Investigation of language relationships provides an important basis for reconstructing the history of peoples, when no written records exist.

The Nilotes are classified into three branches on the basis of their language and habitat: the River-lake Nilotes, the Plains Nilotes, and the Highlands Nilotes. I have chosen one ethnic group to represent each subdivision: the Luo, the Maasai, and the Kalenjin, respectively. All these peoples entered East African from the north or northwest over a period of many centuries.

The Luo people of western Kenya speak a language which is somewhat different from that of their relatives in the Sudan, Uganda, or the Congo. About two million Luo now live in Kenya, and are outnumbered only by the Kikuyu. Originally they were pastoral, with no attachment to the land, but when the British froze the boundaries in Kenya, as in their other colonies, each family developed close ties to the land of its ancestors.

The most famous of the East African pastoralists are the Maasai, perhaps because they have adhered most rigidly to the customs and culture of their forebears. Related to them are the Arusha people of northern Tanzania. Culturally the Maasai have similarities to the Nandi, a branch of the Kalenjin. This last category also includes the Kipsigis, Keiyo, Marakwet, Pokot, Sabaot, Terik, and Tugen, all residing in western Kenya.

The Bantu-speaking peoples migrated from the Cameroon region in the west over a period of many centuries and are dispersed throughout all parts of East Africa, as well as most of southern Africa. They have been iron-working agriculturalists for perhaps two thousand years. I shall deal with the Kikuyu, the Kamba and Taita of southern Kenya, the Chagga of Mt. Kilimanjaro and their neighbors, the Shambaa, in Tanzania, the Ganda of Uganda, and the Swahili along the coast.

Trend Towards Assimilation

Traditionally the Nilotes are associated with cattle, while the Bantu are considered agricultural peoples. Economies have changed over the years, and there are many exceptions to the generalization. The Nilotic Maasai people are famed as nomadic pastoralists, and their wealth consists entirely of cattle. Yet the Maasai people of Arusha, in northern Tanzania, have been agriculturalists for over a century, while retaining their original language and culture patterns.

Mr. Kimalel told me that his people, of the Nandi group, a subdivision of the Kalenjin, had formerly been pastoral, but were now attached to the land. His grandfather had acquired great wealth in cattle. Now the family owns over two thousand acres of farmland in western Kenya. They cultivate with the most modern equipment, and graze large herds of European grade cattle, which are a cross between native cows and European bulls.

On the other hand, some of the Bantu-speaking peoples, while retaining their attachment to the land, have also adopted the "cattle complex." The Kikuyu, for example, are just as devoted to their cattle as are their Nilotic neighbors.

Even where there were once distinct castes in a society, differences have gradually disappeared. Rwanda had three main groups. The famed Tutsi (or Watusi), the tallest people in the world, cattle-owning Nilotic aristocrats, for centuries dominated the Hutu, Bantu agriculturalists, and the Twa, a very short people living in a state of near-serfdom. In spite of their position of dominance, the Tutsi accepted the Bantu language of the Hutu as their own. Now all the people of the country speak a common Bantu language, Kinyarwanda (language of Rwanda).

The system of age-grades and the rite of circumcision as a part of the initiation into adulthood were acquired through contact with the Cushitic people, who probably migrated from Ethiopia. These customs were adopted by many Bantu and Nilotic peoples in Kenya and northern Tanzania.

Economic and Political History

Two thousand years ago Greek sailors were guided down the shores of the Indian Ocean by the *Periplus of the Erythrean Sea*, a book that said of the ancient market town of Rhapta, near the present site of Dar es Salaam: "the men are very tall, and intermarry with Arabs." All along the coast there arose small, flourishing city states, collectively known to the Arabs as the land of the Zanj. The mixture of Arab and Persian with the Bantu-speaking peoples produced the unique Swahili culture.

In the National Museum of Tanzania one can see some of the copper coins minted by the rulers of Kilwa, a city state on the Tanzanian coast. The forty-two Sultans described in the Kilwa Chronicles reigned in the period 1285–1493 A.D. The Portuguese explorer, Duarte Barbosa, wrote in 1517 that the wealth of Kilwa was due to the monopoly of the gold trade from Sofala, further south on the East Coast. This gold, mined inland, was exchanged for cotton and silk cloth and for beads at the Sofala market.

Ivory was another product in great demand at that time. We know little about the means of conveying goods from the inland regions to the coast, before 1800. Probably the gold and ivory that contributed so much to the enormous wealth of Indian princes was bartered from tribe to tribe, until it was purchased by Swahili, Arab, or Indian merchants at the coast.

Soon after 1800, the pattern changed. Caravans traveled hundreds of miles to the large markets, there to trade in ivory, iron implements, cloth, baskets, and livestock. The chief middlemen in the long-distance trade between the interior and the coast were the Kamba. The main item of their commerce was ivory. In exchange, the Africans received cloth, wire, beads, and some firearms.

By mid-century the Swahili and Arabs were sending their own caravans far inland. Ivory and slaves constituted their most precious "cargo." The two "commodities" were inseparable. After carrying the "white gold" over vast distances on their heads, the Africans were then sold into slavery. A hundred thousand human beings died every year along the ivory trails from inland Africa, exhausted by disease, hunger and overwork. In the hunt for elephants, thousands more Africans were slain, their homes, fields and villages destroyed by European and American guns and gunpowder. On the wall of the Museum of the Holy Ghost Fathers Mission in Bagamoyo, Tanzania, is a map showing the ever increasing penetration of the slavers into inland Africa. By the late nineteenth century, the slave caravans were handing over to the local chiefs the *hongo*, or safe transit tax. Payment was in cowries, at the rate of 200 to the Indian rupee, the official currency at the time.

The political organizations of the inland peoples varied from state-lessness to highly developed centralized kingdoms. They were generally correlated with the fertility of the land and its ability to support an expanding population.

The Eastern Bantu and Nilotic peoples retained a well-ordered system of authority, based on allegiance to their lands or their herds of livestock, and administered by the council of elders of each local community. One person might gain sufficient respect from his fellows to become a "first among equals"; there were few chiefs or kings. Further west around the Great Lakes, several powerful kingdoms emerged. Among the foremost was Buganda, the kingdom of the Ganda. Originally the kings were the elected leaders of the councils of chiefs. Gradually they became strong enough to collect tribute from the lesser chiefs. Just as in the kingdoms of England and France, the Ganda rulers became increasingly despotic. The country was divided into units for the purpose of civil and military administration. A well-developed system of roads enabled the central government to maintain efficient communication with the provinces. By the late nineteenth century, Britain had established a "protectorate" over Buganda and soon extended it to the other parts of present Uganda. Mutesa II, the ruling Kabaka, or king, last of a six-hundred year dynasty, was exiled to England in 1953 for failure to cooperate with the British. He returned subsequently, but was again exiled in 1966 by the independent Ugandan government, which abolished traditional kingdoms the following year.

British domination of East Africa produced violent disruptions in the land that later became Kenya. British and South African colonists had no qualms about displacing the native population from the desirable Highland regions. Masses of people were resettled from their fertile ancestral lands to the unproductive areas. Boundaries were drawn, cutting right across the territories of ethnic groups with no regard to these peoples. Kenya gained independence in 1963 after the violent struggles of the famous Mau Mau uprising, led by *Mzee* (Old Man), Jomo Kenyatta, now the president of the country.

The present state of Tanzania was formerly a German "protectorate," German East Africa, established in 1885. In 1920, after World War I, the British took over from the defeated Germans and changed the name to Tanganyika. The country won its independence in 1961, and three years later joined with the "spice island" of Zanzibar to become Tanzania, under the presidency of Julius Nyerere, affectionately called *Mwalimu*, "Teacher."

CHAPTER 21
SPOKEN AND GESTURE COUNTING

David Zarembka, an American volunteer teacher, lived and worked among the refugees from Rwanda in western Tanzania. At the market one day he asked the price of an item. "One shilling," replied the merchant, with a wave of his hand. David offered the shilling, but the merchant refused to sell the article. Again he waved his hand, with two fingers outstretched. At last David understood that the price was one shilling, twenty cents. The people of Rwanda, accustomed to the Belgian franc, convert ten cents in Tanzanian coinage to one franc. The merchant expressed the price as one shilling and two francs, indicated by waving two fingers.

Gestures are used either for emphasis or to replace the spoken word. The Hima (Nkole) cattle-owner may say: "I have cattle tens so many," as he indicates with his fingers the number of tens of cattle he possesses. The gesture may be an attempt to circumvent the taboo on counting living creatures. In the market place, gesture counting overcomes the language barrier between people of different backgrounds. In a different context, the fingers are used as an aid in mental arithmetic. Rwanda secondary school students told me that even uneducated (unschooled) people can tick off a multiplication problem on their fingers to calculate the cost of several units of an item. To illustrate, one boy pointed to his fingers in succession as he said, "Twenty-five, fifty, seventy-five. . . ."

Counting to Nine in Bantu Languages

In her study of African numeration, Marianne Schmidl devoted particular attention to the Bantu languages. The following excerpts describe both spoken and gesture numeration in East Africa in the early years of the twentieth century (pages 171–175):

The number words for two through five are as already described [See Chapter 4]. In some languages the words for six through nine are formed by composition with five: 6 = 5 + 1, etc., but in others, the names are based [on a subtractive principle or] on the principle of two approximately equal terms. I will cite the Shambaa language [of northeast Tanzania] as an example:

6 = mutandatu = ntatu na ntatu = 3 + 3.
7 = *mufungate* = *funga ntatu* = bind 3 (fingers); from the ten fingers seven remain; therefore a subtractive formation.

8 = *munane* = *ne na ne* = 4 + 4.
9 = *kenda*, an alien word, used here, as in other languages,
without the class prefix.

With the exception of the widespread dispersion of Swahili, this
system of numeration can claim a territory which extends from the
mouth of the Tana on the northern boundary of the Bantu [in Kenya],
to Lake Nyanza, including the whole central plateau, Burundi and
Rwanda, and extending to the other side of the central African Rift
Valley. The southeast boundary forms a line from Bagamoyo to the
point where latitude ten degrees South meets the Nyasa, from which
the Ngoni river flows southward. It also occurs among the Fipa, as well
as the Konde north of the Nyasa.

Here we are dealing with a number construction based on
the principle of two approximately equal addends, corresponding
to the finger gestures, where not only is 6 represented by 3 + 3
and 8 by 4 + 4, but 7 is 4 + 3 and 9 is 5 + 4.

It is not surprising that in the border areas one finds that these
constructions alternate with the use of five as a base. The Tabwa and
the Konde count according to both methods. Often certain expressions of one number system are adopted in the other, as we have seen
in the case of 7 = *fungate*, so that the fact that this number word, like
kenda (= 9), requires no prefix in most languages, perhaps points to
its special position in the number sequence. On the other hand many
expressions from the Shambaa numeration system are widely used
in the quinary areas.

When we examine gesture counting, we again find the two systems. The quinary (base 5) is exemplified by the Soga, who indicate 6
by holding the left index finger near the closed right hand; for 7 they
grasp the left index and middle fingers, etc. To indicate 6 to 9, the
Chagga (of Moshi) encircle the fingers of the right hand, starting with
the little finger, with the whole left hand; similarly the Tete, but instead of grasping the fingers, they cross the appropriate ones with
the left thumb. Frequently the bent fingers rather than the extended
fingers are counted, as in the case of the Tete for 1 to 5, as well as the
Hehe.

Far more interesting than this quinary system is the method of
finger gestures based on the principle of two approximately equal
terms. Again I cite the Shambaa. They indicate:

1 by extending the index finger of the right hand.
2 by extending the index and middle fingers of the right hand,
spread out.

3   by extending the three outer fingers of the right hand, spread out.
4   by separating the four outer fingers of the right hand, so that the little and ring fingers touch on one side and the middle and index fingers on the other side.
5   by making a fist
6   by extending the three outer fingers of each hand, spread out.
7   by showing four on the right hand and three on the left.
8   by showing four on the right hand and four on the left.
9   by showing five on the right hand and four on the left.

This method of gesture counting has been adopted with only slight variations by numerous tribes in the northeastern language areas. Its characteristics are as follows:

Counting begins with the extended forefinger. The gesture for 4 is usually 2 + 2. Five is the closed fist, frequently with the thumb placed between the middle and ring fingers, to indicate $5 = 1 + 2 + 2$. (For example, the Kinga, Hehe, and Nyaturu.) The numbers 6 through 9 are expressed in the manner discussed in connection with the spoken number words of the Shambaa people. This method of finger counting occurs essentially in the same areas as the analogous system of verbal numeration.

It is natural that counting gestures based on five sometimes were used interchangeably with the aforementioned symbols. Indeed, there are cases in which the numbers of a particular sequence are constructed on several different principles; thus, the Songora express 6, 8, and 9 in the manner of the Shambaa, but 7 is 5 + 2. We also note discrepancies between spoken and gesture counting; the Chagga form their number words as do the Shambaa, while they count with gestures based on five.

The numeration system of the Rundi is either quinary or according to the principle of two approximately equal terms, depending upon the object of the counting. When the Rundi have to count large quantities of beads, which serve as money in all of Burundi, they always count off by fives.

Instead of:		They say:	
$6 = itandatu$	$= 3 + 3$	itano n'umwe	$= 5 + 1$
$7 = indwi$		itano n'iwiri	$= 5 + 2$
$8 = umunane$	$= 4 + 4$	itano n'atatu	$= 5 + 3$
$9 = icienda$		itano n'inne	$= 5 + 4$

Since the Rundi never speak a numeral without making the appropriate gesture, when they count beads, they use only quinary signs, rather than the gestures of the type used by the Shambaa. This is of great significance, since among the most heterogeneous peoples, only a single method of counting off by fives has been observed.

Now I will describe a system of gesture counting which has little resemblance to those already discussed, namely, that of the Ziba near Lake Victoria and the Hima of Ankole. Their gestures are:

1 through 3   like the Shambaa
4 and 8   place the tips of the middle finger and thumb together, and click the ring finger against them either twice or four times.
5   extend the lower joints of all five fingers, and bend the upper joints.
6   raise the ring, middle and index fingers.
7   raise the small, middle and index fingers.
9   raise the outer fingers.

Although the representation of 4 and 8 is similar to that of the Shambaa, the expressions for 6 and 7 tend toward a quinary base. It is reported, further, that for 6 the thumb and little finger are clicked, for 7 the thumb and middle finger, and for 8 the thumb and ring finger, so that the thumb represents the whole hand—that is, 5.

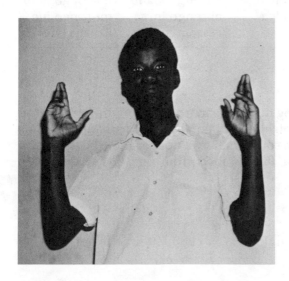

Figure 21–1 Gesture for "eight," Rwanda and western Tanzania. Only motion pictures can do justice to this gesture. The forefinger of each hand is snapped sharply against the ring finger, and comes to rest on the middle finger. "Four" is denoted by a similar motion with just one hand.

The noteworthy feature of the Eastern Bantu number words for six through nine is that they may be formed according to two entirely different principles, or a combination of both. In most East African Bantu languages, 6 = 3 + 3 and 8 = 4 + 4. But *mufungate* (seven) derives from the expression meaning "bind three fingers," or 7 = 10 − 3. Similarly kenda (nine) means "take one"; that is, 9 = 10 − 1. Thus the words for six and eight are based on the principle of two equal terms, while seven and nine are formed by subtraction from ten (Table 9).

	Swahili	Kamba	Taita	Rwanda
1	moja	imwe	imweri	mwe
2	mbili	ili	iwi	bili
3	tatu	itatu	idadu	tatu
4	nne	inya	inya	ne
5	tano	itaano	isanu	tanu
6	sita [Arabic]	thanthatu	irandadu	tandatu
7	saba [Arabic]	muonza	mufungade	rindwi
8	nane	nyaanya	inyanya	munani
9	tisa [Arabic]	keenda	ikenda	icyenda
10	kumi	ikumi	ikumi	cumi
11	kumi na moja	ikumi na imwe	ikumi na imweri	cumi na-mwe
12	kumi na mbili	ikumi na ili	ikumi na iwi	cumi na-bili
20	ishirini	miongo ili	mirongo iwi	ma-kumi a-bili
30	thelathini	miongo itatu	mirongo idadu	mi-rongo itatu
40	arubaini	miongo ina	mirongo inya	mi-rongo ine
100	mia moja	iana yimwe	ighana	ijana

Table 9   Bantu Number Words

It would be interesting to be able to trace the origins of variations in gesture counting among neighboring peoples. The Kamba and the Taita peoples, both of whom live in southern Kenya not distant from the Shambaa, have similar gestures for the numbers from one to five. After five we see:

	Taita	Kamba
6	right fist, extend left thumb (5 + 1)	clasp left little finger (5 + 1)

Taita	Kamba
7 right fist, left thumb and index (5 + 2)	clasp two outer fingers (5 + 2)
8 extend eight fingers, no thumbs (4 + 4)	clasp three outer fingers (5 + 3)
9 clasp left fingers in right hand (5 + 4)	clasp four left fingers (5 + 4)
10 both fists (5 + 5)	both fists (5 + 5)

In both languages the number words for six mean "three and three" and the words for eight mean "four and four."

Figure 21–2   Kamba finger gestures for one to ten.

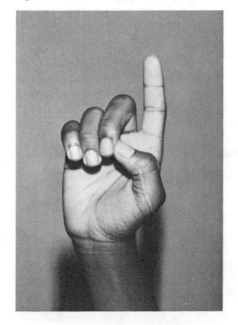

Imye
(one)

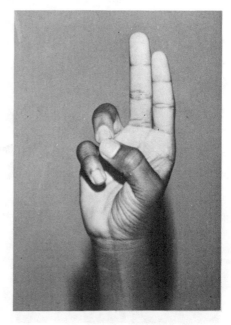

Ili
(two)

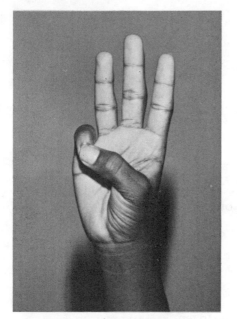

Itatu
(three)

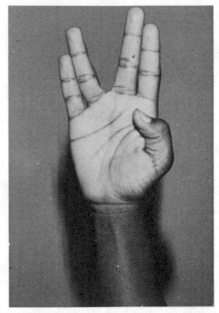

Inya
(four)

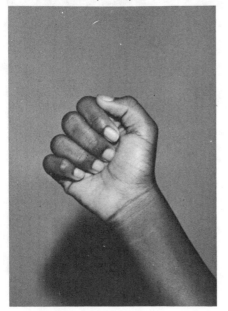

Itaano
(five)

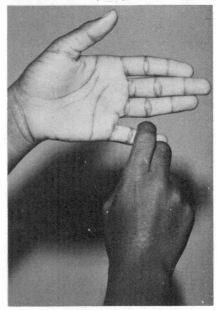

Thanthatu (3 + 3)
(six)

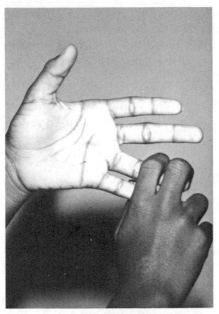

Muonza
(seven)

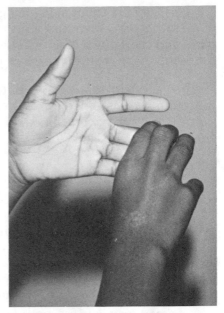

Nyaanya (4 + 4)
(eight)

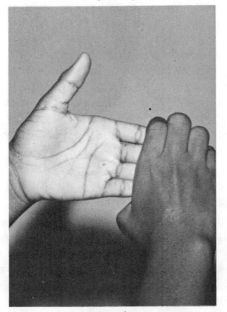

Ikenda
(nine)

Ikumi
(ten)

The Taita children in their games use another system, said to be the numerals in use a hundred years ago. They sing these words to a simple tune, and bend down their fingers with each word. No one knows whether the words are archaic numerals or just nonsense syllables.

Higher Denominations in Bantu Languages

Although *kumi* is generally the word for ten, the higher decades may be formed in combination with *kumi* or with *longo*. In old Swahili, spoken a generation or two ago:

> 20 makumi mawili (2 × 10), etc.
> 90 makumi kenda (9 × 10)

In current usage the Arabic word *tisa* has replaced the Bantu *kenda* as the word for nine, and Arabic words are used for all the multiples of ten. (The Arabic expressions for fractions are also in general use. Instead of the older Bantu formation for one-eighth, *moja juu nane* (one over eight), the Arabic *themuni* is used in modern Swahili.)

Some languages use both *kumi* and *longo* to form ten and its multiples. In the Kamba language (Kikamba) ten is *ikumi*, but twenty is *miongo ili*, and similarly for the higher decades. In the language of the Rwanda people (Kinyarwanda), both ten and twenty are based on *kumi*, but the larger multiples of ten combine with *rongo*, a variant of *longo*.

For centuries the kingdom of Buganda was a dominating power, with an organized system for exacting tribute from all its subjects. Names for large numbers were essential to the economy. In Luganda, the language of the country, ten is *kumi* and one hundred is *kikumi*. There was a special word for ten thousand, *mutwalo*, a unit of cowrie shell currency. The word for ten million means "the uncountable amount." Undaunted, the Ganda continued with "the uncountable amount twice over" to express twenty million.

At the turn of the century the Reverend Hattersley, a missionary, described their ability to reckon (page 174):

> By their own methods the Baganda are very quick at calculating, though they count in a most peculiar manner. A raw peasant, for instance, wishing to count one hundred cowrie shells, will mark off two and two and call them one, and so on five times, when he has got to twenty, and five times of this completes the hundred, but they can invariably tell when one shell is missing in a thousand.

East African Commerce Gestures

The Arab and East African merchants have developed a system of finger gestures that is understood by every trader in the region. Its purpose is to enable the buyer and the seller to reach an agreement in secret, amidst the noise of the open market. It has the additional advantage of overcoming language barriers. Negotiations are conducted under a cloth by touching fingers. First the parties agree on the order of magnitude of the price—whether units, tens, hundreds, or a higher order—or they may specify the denomination of the coin. If the buyer touches the seller's index finger, he indicates "one" of the agreed-upon denomination. Touching the first two fingers signifies two, twenty, or two hundred, and the first three or four fingers have corresponding significance. For five, he touches the whole hand. The little finger alone signifies six, the ring finger alone is seven, the middle finger is eight, the bent index finger is nine, and the thumb means ten. The fractions one-half, one-fourth and one-eighth can be added or subtracted by stroking the appropriate finger from the middle joint to the tip or away from the tip, as the case may be.

To indicate that he offers 350, the buyer touches the thumb three times, then grasps the hand. The operations are very rapid and surprisingly free of error (from Menninger, pp. 212–214).

Nilotic Numeration Systems

To do justice to the finger gestures of some Nilotic peoples, one should have motion pictures with sound. In fact, one African refused to describe for me the gestures of his people, insisting that only photographs could convey them properly.

	Maasai	Kalenjin (Nandi)	Luo
1	nabo	agenge	achiel
2	are	aeng'	ariyo
3	uni	somok	adek
4	onguan	ang'wan	ang'wen
5	imiet	mut	abich
6	ile	lo	auchiel (5 + 1)
7	naapishana	tisap	abiriyo
8	isiet	sisit	aboro
9	naaudo [Kenya] enderuj [Arusha]	sogol	ochiko or ongachiel (10 − 1)
10	tomon	taman	apar
11			apargachiel (10 + 1)
20	tikitam	tiptem	piero ariyo
30	tomoni-uni [Kenya] osom [Arusha]	sosom	piero adek

	Maasai	Kalenjin (Nandi)	Luo
40	artam	artam	piero ang'wen
50	onom	gonom	piero abich
60	ntomoni-ile	tamanwogik lo	piero auchiel
	formerly onom o tomon		
70	ntomoni-naapishana	tamanwogik tisap	piero abiriyo
	formerly onom o tikitam		
100	iip *or* iip nabo	bogol	piero apar achiel (10, 10 times, once)

Table 10   Nilotic Number Words

It is said about the Arusha Maasai, of northern Tanzania, that they rarely give numbers verbally without the accompaniment of finger signs. At times the spoken expression is omitted, and the listener is expected to voice the numbers in response to the observed finger action, to ensure that both agree on the number. The Arusha people expect the listener to indicate by words of affirmation that he is following the speaker at all times, and this practice applies to numbers as well.

The Arusha Maasai count on one hand only, usually the right. Here I rely upon Dr. Gulliver's description. To simplify, I shall name the fingers T, 1, 2, 3 and 4, representing thumb, forefinger, middle finger, ring finger and little finger, respectively.

1	extend 1
2	extend 1 and 2, move them scissor-like
3	ball of 1 on T, tip of 2 resting on 1
4	extend 1 and 2, with 2 resting on 1
5	closed fist, T protruding between 1 and 2
6	tip of T to tip of 3, with several clicks of the nails
7	rub T and 2
8	extend 4 fingers and move them slightly in the plane of the hand
9	T and 1 form a circle, other fingers are extended
10	begin with T and 1 forming a circle, then extend 1 sharply; other fingers are curled up
11–19	gesture for ten, followed by gesture for the units digit
20	the open hand is closed with a snap two or three times

Figure 21–3   Finger gestures of the Arusha Maasai. (Source: Gulliver).

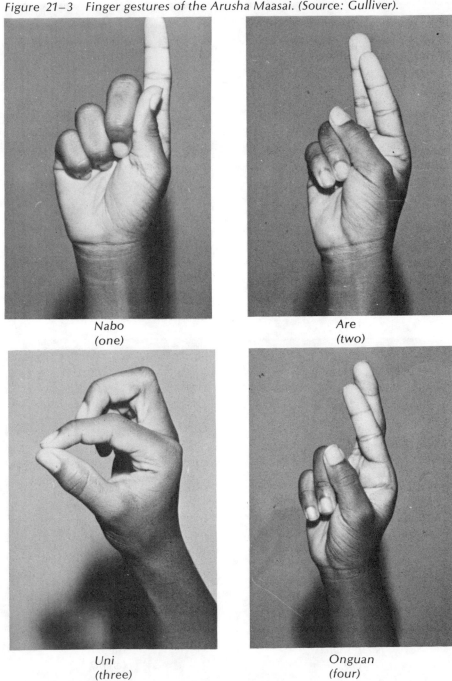

Nabo
(one)

Are
(two)

Uni
(three)

Onguan
(four)

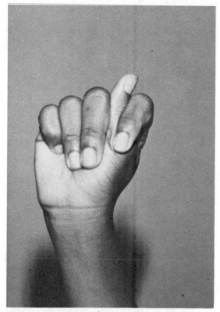

Imiet
(five)

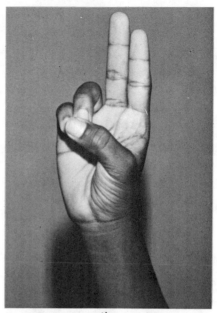

Ile
(six)

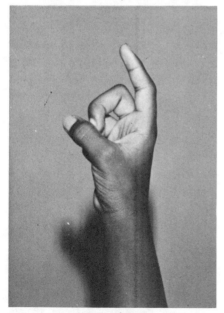

Naapishana
(seven)

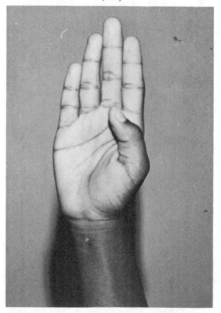

Isiet
(eight)

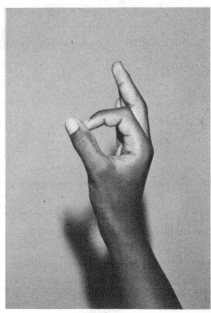

Enderuj
(nine)

Gulliver gives gestures for the principal numbers up to 2000.

Several generations ago the Maasai words for 60 and the higher decades were formed by composition with fifty: sixty was *onom o tomon* (50 and 10). Today the word for sixty means ten times six; seventy, eighty and ninety, too, are expressed as multiples of ten. One thousand is "ten hundreds." There are distinct words up to one million, the "uncountable."

The gestures of the Kalenjin people, as represented by the Nandi, resemble those of the Maasai. The same cannot be said about the number words. Comparing the numerals below ten, only the words for four are similar. The similarity in the words for the higher decades is probably due to borrowing.

The gestures for 1, 2, 4, 5, 7 and 8 are the same as those described for the Arusha. For the other numbers the Kalenjin show:

3	tips of T, 1 and 2 together
6	rub T and 1
9	T and 1 form a circle, with the tip of the index finger of the other hand placed on the juncture of T and 1
10	begin with T and 1 forming a circle, then extend 1 smartly so that it snaps against 2

11–19   gesture for ten, followed by the units digit
20   the open hand is closed with a snap four times
30   T and 2, tips together, shake slightly
40   shake the hand vigorously, with all four fingers extended
50   throw out the closed fist, then open it
over 50   same as for 50

	Gestures	
ETHNIC GROUP	One	Four
Shambaa (N.E. Tanzania)	Extend R1	Extend R1,2; 3,4
Kamba (southern Kenya)	same	same
Taita (southern Kenya)	same	same
Arusha Maasai (N. Tanzania)	same	Extend R1,2, with 2 on 1
Nandi (W. Kenya)	same	same as Arusha
Luo (W. Kenya)	same	same as Shambaa
Tete (Mozambique)	Bend R1	Bend R1,2,3,4
Hima (Nkole) (Uganda)	Extend R1	Snap R1 against Th, extend R1,2,3,4

Table 11   East African Finger Gestures. (Fingers are designated Thumb, 1,2,3,4)

Not all Kalenjin people have gestures for all the numbers. For example, the Marakwet have no gestures for 7 or for 9.

The traditional number system ended with 50. At present it extends to 999, after which Swahili is used for precise numeration. For numbers beyond one thousand, Kalenjin people may use the term *bogol makirar*, meaning "hundreds unreckonable."

Gestures		
Six	Eight	SOURCE
Extend R2,3,4 and L2,3,4	Extend R1,2,3,4 and L1,2,3,4	Schmidl
Clasp L4 with R hand	Clasp L2,3,4, with R hand	Kamba students
R fist, extend L. Th.	same as Shambaa	Williamson
R Th & 3 form circle, nails click	Extend R1,2,3,4 and move them in the plane.	Gulliver
Rub R Th. and R1	same as Arusha	Kimalel
L fist, extend R1	L fist, extend R1,2,3	Ayany
L Th on R4 (extended)	L Th on R2,3,4 (extended)	Schmidl
Extend R1,2,3, snap Th & 4	Extend R1,2,3, snap 1 against 3 to hit 2	AVO and Roscoe

The Luo system of finger gestures is a pure quinary one, and easy to describe. Again the symbols T, 1, 2, 3 and 4 are used to represent the thumb, forefinger, middle finger, ring finger, and little finger respectively. The signs for the first five numbers are performed on the right hand.

1	extend 1
2	extend 1 and 2
3	extend 1, 2 and 3
4	extend four fingers, with 1 and 2 separated from 3 and 4
5	closed right fist
6	closed left fist and extend right 1
7	closed left fist and extend right 1 and 2
8	closed left fist and extend right 1, 2 and 3
9	closed left fist, and right hand as for 4
10	two closed fists
11–19	two closed fists, followed by the appropriate gesture for the digit

There are no gestures for the numbers beyond 19. Sticks may be used as an aid in counting, with no particular grouping.

Luo number words are also based on five, with ten as a secondary base. Six and seven are formed in composition with five; nine might be 10 − 1, or an independent word, and ten is an independent word. Beyond ten the number words are formed by addition of the primary digits to the multiples of ten. Mr. Ayany told me there are number words up to a million. After 1000 (*gana*) the words are composed with *gana*.

Counting Taboos

"How many children are present today?" Marjie Spiers, an American volunteer teacher, asked her class of Kikuyu primary school children. "Twenty-five," came the answer. Marjie counted out twenty-six sheets of paper for the spelling drill. These children circumvented the taboo on counting living things by declaring the number to be one less than the correct one.

A British visitor to a Kikuyu village, early in the twentieth century, asked a mother how many children she had. "Come and see," was her evasive reply. However, he found that Kikuyu boys readily answered questions about the number of their father's wives and of their respective children. Boys will be boys—at any rate, when the adults are not around.

As a youngster, Jomo Kenyatta, the president of Kenya, was required, as were most boys, to watch the family's herds. In his book *Facing Mount Kenya*, he writes (page 102): "Care is taken to teach the boy how to be a good observer and to reckon things by observation without counting them, as counting, especially of sheep, goats, cattle or people, is considered as one of the Gikuyu taboos, *mogiro*, and one which would bring ill-luck to the people or animals counted." Each head of livestock has its individual name, and is recognized by its color, its markings, and by the size and type of its horns.

A Kikuyu boy is given a rigorous training in animal recognition. As a test, two or three herds are mixed, and he is expected to select the animals belonging to his own family. Or several animals are hidden, and the child is asked to inspect the herd and report the missing members. If he makes an error, his tutor asks him to point out that specific animal in the herd. Kenyatta learned his lesson well, and in turn, taught his younger brothers.

This reluctance to count humans and animals for fear they may be harmed is widespread among all the ethnic groups of East Africa. As the herds, numbering perhaps hundreds of cows, goats and sheep, return at night, their keepers scan them to detect missing animals. The languages include numerous words descriptive of the varieties of domesticated animals, based on color, markings, horns, and other features. Even the agreement on the bridewealth a young man is obliged to deliver to his future in-laws may specify the name and description of each animal.

In his comparison of the numeration system of the Arusha Maasai with that of the Turkana, a pastoral people of Nilotic origin, Gulliver declares that the Turkana, even today, have little occasion to use numbers over fifty, while the agricultural Arusha Maasai have now become accustomed to counting large numbers of shillings. Although the Turkana have far larger herds than the Arusha Maasai, they do not need to count their animals, for they know each one by sight. "They no more think of counting their herds in the pastures or the kraal than they think of counting the members of their extended family at a ceremonial meeting" (page 262).

Special Numbers

Are certain numbers considered lucky, and others unlucky? Europeans visiting Africa early in the twentieth century wrote of a widespread belief in the special import of various numbers. Hollis, for example, wrote about divining among the Nandi, a branch of the Kalenjin peoples. For the diviner, the lucky numbers were 2, 3, 5, 8 and 10, especially 3 and 5. On the other hand, 1 is the most unlucky, followed by 6, 7 and 9, and 4 was the least unlucky of the group. Dr. Kipkorir was not familiar with this belief, adding that we must "read Hollis with a pinch of salt." He knew of no Kalenjin word meaning "lucky." I was told that the Luo, also, have no lucky or unlucky numbers.

Obviously there are discrepancies in the reports about the significance of certain numbers! A European, with his particular culture bias, educated in the western scientific tradition, might give an interpretation to a given phenomenon that would differ from that of an African brought up in the traditions of his people. A third interpretation might come from an African who has had a western education. He may have forgotten or never known the ancestral beliefs, or he may regard them as quaint anachronisms.

I shall present the information mainly as reported by European observers and confirmed whenever possible by present-day Africans. Dr. Raum, who grew up among the Chagga, wrote in 1938 that they call an odd number "the number without a companion," and often consider it unlucky. This has been confirmed by young people I interviewed. Kamba students say that such a number is mwa. Children are advised not to walk in groups of three, five, seven, or nine. Bantu-speaking secondary school boys from the Lakes region verified the existence of the belief that even numbers are lucky and odd numbers unlucky.

Lindblom, in his reports of the Kamba people of Kenya, wrote: "Odd numbers are generally considered disastrous or at least unlucky. . . . The number seven is the most prominent of all the numbers. This seems to be the case over large parts of East Africa as well. . . . Among the Akamba, seven is found both as a good and evil number, in most cases, however, it

seems to be bad" (pages 306–307). Lindblom attributes the special importance of seven to the influence of Islam, and its prominence can indeed be
traced to pre-Biblical times, as evidenced, for example, in the seven-day
week.

Among the Kamba people, noted Lindblom, the cattle in pasture
are never watched by the same cowherd for more than six days at a time.
Perhaps this was a form of labor protection, like the five-day week! A
farmer safeguards his sugar canes from thieves by placing seven porcupine
quills in the stalks. Seven is important in oaths and incantations. On the
other hand, seven is also associated with auspicious occasions—the seven
days of the circumcision festivities, or the cracking of a whip seven times
to bring good fortune to the elephant hunters.

The Rev. Dr. Mbiti, himself a Kamba, wrote me: "The only unlucky
number [among the Kamba] is seven. But perhaps the word unlucky is not
the correct one, because people simply don't like the number seven, as
they consider it to be bad, although they do not specify in what ways."

The number seven appears to have special significance among the
Bantu-speaking peoples living near the coast. Raum wrote that the Kikuyu
have no number name for seven, but use a non-number word instead.
[The word for seven is *mugwanja*.] Youths, continued Dr. Raum, are forbidden to travel in groups of seven, and people do not divide anything into
seven shares. A school child, confronted with the arithmetic problem: "A
man has eaten seven fishes. If he eats three more, how many will he have
eaten?" might very well respond: "He'll be dead by tomorrow for eating an
unlucky number!" In his informative book *Arithmetic in Africa*, Raum
suggests that teachers take advantage of the taboo on seven by having the
child say "whizzbuzz" whenever he comes to a multiple of seven or a combination with seven.

Among the Kikuyu the number seven is crucial in casting spells.
Seven curses may be invoked, each guided in its flight to the home of the
victim by the alignment of one of seven sticks. Kenyatta relates, from his
boyhood, the procedure a man must follow when he vows to devote his
life to love. He must first denounce all property, such as goats, sheep and
cattle. Then the magician takes him to a lonely spot where the hyenas live,
there to invoke the appropriate spirits, chanting the invocation seven times
as he swings a magical bag over the love-seeker's head. Then he beats the
suppliant seven times on the buttocks with the magical bag, as he asks the
ritual questions. This ceremony is followed by the drinking of a love potion
and the anointment of the body. Now the love-seeker is prepared to
"devote his life to love-making."

To the magician, the number seven is of great import. In the National
Museum in Nairobi is a display of the Kikuyu magician's gourd and various

small objects used as counters—seeds, stones, and items of historical or magical significance. When he is consulted, he pours the counters out on the ground and groups them into heaps of ten, combining the piles to form twenties, fifties, and hundreds. The crucial number is the remainder. If seven objects are left over after the grouping, this may signify death. A remainder of three means that it is proper to proceed with the casting of lots to answer specific questions—is it the proper time to undertake a journey, or will a sick person recover? In divination, the numbers three and five are considered fortunate, in spite of the general fear of odd numbers. The young people to whom I spoke were familiar with the procedure.

To the Ganda people of Uganda, nine is the most significant number. The Kabaka of Buganda played a leading role at the national festivals in making an offering of nine men, nine women, nine head of cattle, nine goats, nine fowls, nine loads of bark-cloth, and nine loads of cowrie shells.

When the Ganda magician is consulted, he throws onto a mat nine strips of leather decorated with cowrie shells. The pattern in which they fall determines the decision. If two lie side by side, with a third strip across them, this is considered a bad omen, and foretells death in the case of illness. Other methods of divination are the casting of nine sticks into a pot of water or nine coffee beans onto a surface. If they form an even number of groups, the outcome is unfavorable, and the formation of just two groups signifies death. Obviously the occultist has some leeway in deciding what constitutes a group! Divination is also carried out by the inspection of the specks on chicken entrails—again, an odd number is a favorable sign.

Certain numbers are associated with childbirth. The Kikuyu father cuts four sugar canes to celebrate the birth of a girl, and the mother and child are secluded for four days. In the case of a boy, the number is five. The juice of the sugar cane is fed to both the mother and child, to ensure their health.

The Taita people of Kenya also associate a number with each new-born child. The father kills a goat and tries to foretell the child's future by examining the entrails. The number given to the child expresses his relationship to the extended family, and henceforth he is referred to as a member of, for example, the sixth group; "the one belonging to six" is one of his given names.

Nature is the Time-Keeper

Rodah Zarembka looked at her watch after the evening meal. "Two-thirty," she announced. I looked at her in surprise. "I mean eight-thirty," she corrected herself, to adjust to my way of thinking about time. Actually her watch did read 8:30, but East Africans mentally reverse the direction of the hour hand. Thus 8:30 is read as 2:30.

Figure 23–1 According to East African time, 8:30 is read as 2:30.

Kenya and Uganda lie directly across the equator, and sunrise occurs at the same time of day from one year's end to the next. Blessed with such a reliable natural timekeeper, the people of East Africa start their day at sunrise. One o'clock corresponds to 7:00 A.M. standard time. Although most official clocks conform to standard time, people automatically translate into East African or "Swahili" time. An exception is the island of Zanzibar, the romantic "Isle of Cloves," off the Tanzanian coast. Here the great clocks in the tower of the House of Wonders and in the new sports stadium are set in the true Swahili tradition. More and more, however, the Africans are conforming to Western conventions.

Of course, the farmer does not look at the clock to see when it is time to take the cattle to pasture. The sun, since time immemorial, has been the guide to the day's activities. Throughout the year the shadows remain essentially of equal length at any particular time of the day, and the African is adept at recognizing the proper time for the daily rounds.

Frequently the periods of the day are named according to the principal activities. The Ankole, cattle-herders of Uganda, divide the day approximately as follows, reports Dr. Mbiti (*African Religions and Philosophy*, page 20):

6 a.m. is milking time (*akasheshe*).

12 noon is time for cattle and people to take rest (*bari omubirago*), since, after milking the cattle, the herdsmen drive them out to the pasture grounds and by noon when the sun is hot, both herdsmen and cattle need some rest.

1 p.m. is the time to draw water (*baaza ahamaziba*), from the wells or the rivers, before cattle are driven there to drink (when they would pollute it, or would be a hindrance to those drawing and carrying the water).

2 p.m. is the time for cattle to drink (*amasyo niganywa*), and the herdsmen drive them to the watering places.

3 p.m. is the time when cattle leave their watering places and start grazing again (*amasyo nigakuka*).

5 p.m. is the time when the cattle return home (*ente niitaha*), being driven by the herdsmen.

6 p.m. is the time when the cattle enter their kraals or sleeping places (*ente zaataha*).

7 p.m. is milking time again, before the cattle sleep; and this really closes the day.

For the Kipsigis people of the western Kenya highlands, the day is divided into dawn, sunrise, early morning (called "the herds may go for early grazing before being milked"), noon, afternoon ("the sun is bending"), sunset, evening, night, middle of the night (about 9:00 P.M. to 4:00 A.M.), and the "stars are waning." Dr. Kipkorir's name was derived from the words *Kerir Kore*, indicating that he was born at the time of the waning of the stars.

Most Africans scarcely recall the traditional calendars of their ancestors. Dundas wrote, in 1926, that older people among the Chagga could recollect a twelve (sometimes thirteen) month calendar based on a lunar month of four unequal weeks, corresponding to the phases of the moon. The reckoning started with the moon's disappearance, and the divisions of the month consisted of two weeks of ten days each and two of four or five days. Each day was believed to have a particular significance— to bring good or bad fortune to men, or to women, or to cattle, or to the wild beasts.

In most cultures the lunar month is traditional; indeed, the same word is often used to mean both "month" and "moon." Since the moon has a cycle of approximately twenty-nine and a half days, the solar year, based on the rotation of the earth around the sun, consists of twelve lunar months plus about eleven days. Generally, the African peoples attempted to reconcile the lunar month with the solar year. Many folk reckon the new year from the first rains. They assume the year to have twelve months, and then make allowance for the days of waiting until the first heavy rains fall. The names of the months may describe the activities of people in connection with their principal occupation—agriculture or cattle-herding—as related to the weather, or may be merely ordinal numbers.

The Islamic calendar, by which Muslims reckon their holidays, is based on the lunar month, and the first day of the year retrogresses through the seasons. It is believed that the Chagga calendar originally had thirteen months, so that the first month occurred later each year. Note that the names for the last four months of our year, September, October, November and December, have the literal meanings "seventh," "eighth," "ninth" and "tenth," reflecting a period of the Roman calendar when the end of the year fell two months later than in our present calendar.

Modern speakers of Swahili use variants of the English names of the months. There are some attempts to continue the use of the Islamic calendar in areas where Muslims constitute a large part of the population, but few people use it except to identify Ramadan, the month of fasting.

The seven-day week is in general use in Africa, the result of Christian or Muslim influence. In Swahili, Saturday is called *Juma mosi*, the "first day" after Friday, the major Muslim day of prayer. But in many other languages Monday is named "first day," meaning first day of work after Sunday. The Luo call Sunday *Odira*, "the day of no work," while the Teso of Uganda name this day of rest *Sabiti*, probably derived from "Sabbath."

A person living close to nature learns to recognize the signs of the approaching rains—the new growth on the euphorbia trees, the appearance of the Pleiades star cluster. The Kamba are able to use the direction of the sun at different periods of the year to calculate the arrival of the rainy or dry seasons. The farmer sets up an observatory on an open, level plot of land, drawing lines from a chosen reference point to several isolated trees. He observes the sun's furthest advance to the north and to the south, and can then predict the seasons based on the sun's position.

Near the coast, people enjoy two harvests a year, since there are two rainy seasons, the big rains from March through June, and the short rains from October through December. Further inland there is but one rainy season.

Cycling and Linear Age Grade Systems

"It happened at the time of the famine of rice" refers to the year 1898–99, when the Kamba had little to eat except the rice distributed by the government and by the missionaries. Events are recalled in connection with some outstanding occurrence of the past. Among the Kikuyu, at the time of initiation into adulthood, the age group assumes a name based upon a noteworthy event—"the age when the Europeans introduced syphilis" or "the age of the uprising against the Somalis." Thus the history of a preliterate people remained alive from generation to generation. The elders related these events to the young people as they sat around the fire in the evening. In this way the tribe perpetuated the feeling of solidarity and collectivisim over the centuries.

In most East African societies the system of age grades is the outstanding form of organization, cutting across both family relationships and governmental institutions to form a closely-knit society based on age and place of residence. The age grade system may be of the "cycling" or of the "linear" type; both methods furnish an accurate method of dating events in history in societies where no written records are kept.

The Karimojong of Uganda have four cycling age grades, named, respectively, the Zebras, the Mountains, the Gazelles and the Lions, each having a span of a generation, about twenty-five years. Thus the complete cycle has a duration of about a hundred years. Only two are considered active at any one time, a junior grade, for example, the "Mountains" grade, which is still recruiting adolescent boys, and the senior grade named "Zebras," to which their fathers belong. Each of these generation groupings is further subdivided into five age sets, with different responsibilities and status levels. Every five years a new age set is opened to membership by initiation, until the junior grade, the "Mountains," has acquired its full complement of five age sets, covering about a generation. At this time the members of this grade are promoted to the status of seniors, and for the next twenty-five years they form the governing apparatus of this stateless society. The members of the "Zebras" now formally constitute their grandchildren as a new age grade, the "Gazelles," and then retire from active life (Figure 23–2).

Among the Kamba, the Kikuyu, the Luo and the Maasai the system is linear, each new grade advancing at regular intervals into the next status level. The Arusha Maasai have two linear streams of age groups. In about the twelfth year of the cycle of a particular set, the next succeeding set begins its cycle. The Arusha recognize two principal grades, the "young men," or Olmurran, and the "elders," called Olmoruo. Prior to the first age grade there is a period of youth. During the period of junior murran-

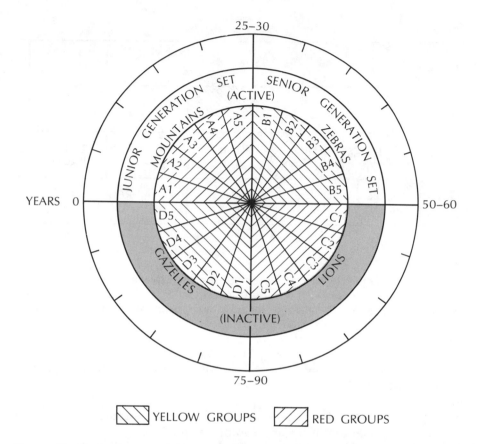

Figure 23–2 Cycling age-grade system of the Karimojong of Uganda. The Karimojong recognize four fixed generation sets, encompassing a total span of 100 to 120 years. Each generation set is subdivided into five age sets. The Gazelles and Zebras, who are called "yellow" because of their brass ornaments, are associated in a grandfather-grandson relationship. A similar relationship exists between the "red" generation sets, the Lions and the Mountains, whose ornaments are made of copper. In this diagram only the Zebras and the Mountains are active, the former in a position of authority and the latter of obedience. (Source: Dyson-Hudson.)

hood, boys are circumcised in three streams, separated by intervals of about two years. The ritual of circumcision is the means of initiation into the age group. This initiation ceremony is followed by a learning period of several years as preparation for promotion to senior murranhood.

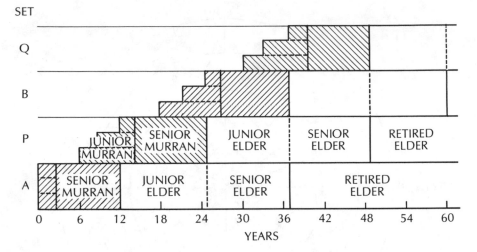

Figure 23–3 Linear age-grade system of the Arusha Maasai of northern Tan-zania. The members of Set B are the wards of Set A, and the members of Set Q are the wards of Set P. Each age set takes an appropriate name, by which it is known thereafter. (Source: Gulliver.)

The age-set cycle has the following stages (the number of years is only approximate) (Figure 23–3):

Years 1– 6: from the close of the circumcision period of the previous set to the resumption of circumcision.
Years 6–14: junior young men (divided into three sections)
Years 14–25: senior young men
Years 25–37: junior elders
Years 37–49: senior elders
After 49: retirement

Tempo is a fourteen-year old of the Arusha Maasai. He has just undergone initiation by circumcision, and is now a junior murran, a member of the same age group as his older brothers, one of whom had been initiated three years before, and the other five years earlier than Tempo. They, however, belong to different sections within the age group. His younger cousin must wait six years before he can be initiated into the next age group, since circumcisions are not performed during this interval. Tempo spends most of his time with the young men of his own circumcision section, learning the rights and obligations of murranhood.

About two years after his initiation, his entire age group, as well as comparable groups in nearby villages, are constituted as an age set and promoted by appropriate ceremonies to the status of senior murranhood. In the old days the young men took pride in killing lions, organizing cattle raids, and defending their herds from marauders. Today life is tamer.

When he is about twenty-eight years of age, Tempo and the members of his age set will be promoted to the status of elders. This position they will enjoy for a period of about twenty-four years, after which they will retire from active community life.

As an old man, Tempo will be held in high esteem by the Maasai community, honored for his wisdom and experience. He may be called upon to render wise counsel in a dispute. He will relate to the young people the events of his life, associating them with the appropriate stage in the cycle of his age grade—the invasion of locusts in the year after his initiation, the terrible drought in the third year after his promotion to the status of *Olmoruo*.

CHAPTER 24
WEALTH MEANS CATTLE—
AND OTHER LIVESTOCK

For the East African, cattle represent security, pleasure, and emotional satisfaction. Each animal in the herd has a name, a recognized personality, and perhaps a poem in its praise. The owner would no more think of slaughtering it than he would his own son. The exception is made for occasional ceremonial meat feasts on the part of the wealthy. The African does not need to brand his cattle, nor does he have difficulty in knowing which animals have been vaccinated; the veterinary assistant merely records the name of the animal in the book, next to the medical information. In many languages there is a large vocabulary to describe cattle solely on the basis of hide markings.

A man's wealth is measured by the number of cattle he owns. It is a matter of prestige for a man to acquire as large a herd as possible. The cattle are useful in supplying milk, and for their hides when they die. Among some Nilotic peoples the blood taken from live animals forms part of the diet. But mainly they are useful as bridewealth, and serve as a measure of the owner's prestige.

In the western kingdoms of Rwanda and Ankole (now part of Uganda), power and privilege rested in the cattle-owing minority, while the majority cultivated the land and cared for the herds. The relationship of the two classes was similar to that of the nobles and the serfs of feudal Europe. The king of Ankole protected all the cattle in the land in return for military service and tribute from his clients. In turn, the agricultural serfs were obliged to contribute produce and labor power to their pastoral lords.

Although cattle were of great importance to the individual owners, their quality left much to be desired. "From the economic point of view the present breed of cattle reared by the Gikuyu is very poor, and it would be a great advancement if the Government could help the people to secure a few good bulls for breeding, and gradually replace the inferior types of cattle with better ones" (Kenyatta, page 65). Kenyatta's advice, written over thirty years ago, is repeated today. In August, 1970, a Kenya government official "appealed to cattle owners in the district to sell their present poor Zebu cattle and go in for graded cattle. . . . Many children suffered from kwashiakor as a result of poor feeding. As this disease was caused by lack of milk, if grade cattle were introduced there would be enough for the children" (*Daily Nation*, Nairobi, August 14, 1970). When David Zarembka was a volunteer teacher in western Tanzania, he observed that twenty cows, the current brideprice in his district, yielded hardly enough milk for the evening tea.

Trade

People who rely entirely on their livestock depend on trade to obtain the necessities of life. By the year 1800 a considerable trade had developed among the eastern Bantu and Nilotic tribes. The Kikuyu supplied food to the Maasai, and received livestock in return. Even in recent times the Maasai were receiving spears, swords, tobacco, gourds, and red ochre from their Kikuyu neighbors, in exchange for sheep.

Figure 24–1   Market at Lushoto, northeast Tanzania. The women wear colorful patterned khange cloths.

Local trade was usually conducted on the basis of barter, at markets held at regular intervals, on a rotating schedule. As many as five thousand people might assemble in one market. These markets still exist today in many areas. The woman with several earthenware pots to sell would like to augment her store of wire ornaments. However, the buyer of pots has a surplus of grain. Exchange of items continues until each person is satisfied.

Certain goods have fixed prices among the Kikuyu. The blacksmith's charge to manufacture a spear to order is a healthy goat. The buyer brings the materials, along with a large calabash of beer to clinch the bargain and put the blacksmith in a pleasant frame of mind.

The Kikuyu were typical of many folk in their use of sheep and goats as standard currency, both before the introduction of a European monetary system as well as a long time afterwards. "They would argue, saying that money is not a good investment, for one shilling does not bear another shilling, whereas a sheep or goat does" (Kenyatta, page 66). Money of itself had practically no function within the country. Even the man who worked for wages and had saved a few shillings would quickly convert the cash into livestock.

Although the actual payment might be made in other commodities, the goat was the measure of value. For an ivory tusk quoted at ten goats, the buyer with a stock of surplus sheep delivers the equivalent number of these animals. Early in the twentieth century the exchange ratio of goats to cattle was thirteen to one in Kikuyu country.

Currency

From early times, cowrie shell currency has been used in Buganda. Before the Arab traders penetrated the country, ivory discs were manufactured by the king's agents and given a value of one hundred cowries. A blue bead also served as a medium of exchange; its value, too, was one hundred cowries.

For large-scale transactions in Buganda, the measure of value was the cow. Roscoe gives the following prices in effect early in the century:

> 1 cow = 2500 cowries
> 1 male slave = 1 cow
> 1 female slave = 4 or 5 cows
> 5 goats = 1 cow
> 1 ivory tusk = 1000 cowries

The cowrie shell had not always been so cheap. In an earlier period two cowrie shells could purchase a female slave. It was the introduction of foreign currency, originally the rupee, which brought about the downfall of the cowrie shell. But the lowly shell lives on in the Luganda word *nsimbi*, formerly meaning "cowrie," but now generally used to denote "money."

A popular form of currency around Lake Nyanza was the iron hoe of local manufacture. Hoes and other iron tools provided an imperishable and transportable medium of exchange. Among the Luyia people three chickens went for one hoe, three hoes for one goat, and three goats for a cow. True to East African tradition, the Luyia measured their wealth in herds of cattle. Other acceptable currency objects in the Lakes Region were brass and copper wire, used primarily for decoration of wealthy ladies, finely woven grass bracelets, cloth, and cowrie shells. In the latter half of the nineteenth century beads and ivory gained in importance.

The institution of bridewealth is an economic device for the redistribution of wealth from the bridegroom's family to that of the bride. Some clans set a standard payment; in others, the amount is negotiated in each individual case. Whatever the custom, livestock forms the basis. "Without them a man cannot get a wife, for it is sheep and goats that are given as *roracio* (marriage insurance). If a man has cash money and he wants to get married he must, in the first place, buy cattle or sheep and goats, because the parents of the wife-to-be will not accept cash money as *roracio*. To them coins have very little meaning and have no religious or sentimental associations within the people's customs" (Kenyatta, pages 66–67). In the 1930s the Kikuyu bridewealth averaged about thirty or forty sheep and goats. If cattle were included, a cow was considered the equivalent of ten sheep and goats.

Kamba women command unusually high payments—two cows, two bulls, and forty or fifty smaller animals is typical, more if the suitor is wealthy. The young man expects to receive contributions from members of his extended family and from his agemates; the bride's father uses the payment to settle his debts. Thus wealth, embodied in the herds, is redistributed throughout the community. Today the Kamba may pay the marriage insurance in cash, the amount depending on the clan tradition, the bargaining power of the parties, and the girl's education. An average payment is about $300 in United States currency. The equivalent is twenty head of cattle, regardless of their condition. The condition may, indeed, be deplorable, since many people keep cattle only for prestige and for bridewealth payments.

In parts of the Musoma district of northwest Tanzania, inflation has seriously affected the custom. Two generations ago the bridewealth was merely one goat. Thirty years later it had risen to six cows; now fifteen to twenty cows are commonly demanded, says David Zarembka.

Among the Kalenjin people of western Kenya, the bridewealth is more modest. Not all the Kalenjin people observe the practice. For those who do, the traditional payment is four head of cattle, sometimes five. Each animal is named and described: "We are giving you a cow with certain marks, brown and white, of such-and-such age; she has calved twice, and she has a sweet disposition." The bride's family may refuse a particular animal, and request another.

The Arusha Maasai transfer a specified number of named animals, consisting of four head of cattle and seven smaller animals. The instalments are arranged by the bride's father, and may be spread out over many years. Nowadays an equivalent cash payment may be substituted, but even a cash payment is described by the name of the animal it represents.

Fines

The social organization of the East African peoples requires a constant supply of stock for marriage insurance, payments of fines, sacrifices, magical rites, and purification ceremonies, as well as for meat and clothing. My Tanzanian friend told of his father's opposition to his marriage to a certain young lady. He explained that the girl was a distant cousin, and there was a taboo on such a marriage. "But we'll kill a goat," he shrugged, "and that will make it all right."

African traditional law embodies an elaborate system of fines in livestock for every category of injury committed against other individuals. The criminal is regarded as a person who can be rehabilitated by his family and community, and his punishment is the payment of a specific number of sheep or goats, depending upon the magnitude of the crime. The only exception is murder by witchcraft. This is regarded as a crime against the whole community, and the penalty among the Kikuyu in former times was death by burning.

In the case of a crime, the injured party brings the complaint before the council of elders for settlement. In serious cases, the offender is ostracized. According to Kenyatta, the stigma attached to ostracism is far greater than that of the European punishment by imprisonment. As an instance of Kikuyu justice, a young man who makes a girl pregnant outside of marriage is fined nine sheep or goats, in addition to three big fat sheep for the members of the council. The young man is then ostracized by the young people of his age group. The girl shows her gratitude for their support by providing a feast for all the young people.

SECTION 8
PAST AND FUTURE

Mathematics plays a part in many aspects of African life—in the market-place and at the religious shrine, in children's games and in the secret arts of learned men. But have Africans in the lands south of the Sahara made a contribution to the development of pure mathematics? That is a question I cannot answer, and I have found no one who can. Available evidence permits us only to conjecture about the past. With the development of the new African nations and the rise of native universities staffed by Africans and sympathetic foreigners, the history of Africa is being pieced together. As we learn more about the various cultures and their achievements, we will gain further insight into their contributions to mathematics.

Figure 25–1   Science Building, University of Nairobi, Kenya.

CHAPTER 25
PURE MATHEMATICS IN AFRICA

It is only in the past few years that European investigators have admitted the possibility of African contributions to world culture. Seventy-five years ago, L. L. Conant, in *The Number Concept*, called the Africans "savages." More recent authors have attributed every African accomplishment to the Europeans, to the Arabs, to the Indians, to the Mesopotamians, to ancient Egypt (whose people they classified as Caucasoid in spite of evidence that many Egyptians were dark-skinned; see Diop and Strouhal)—in short, to any source but black Africans themselves.

Now it is recognized that Africa has contributed much to the world; yet I have found no indication of Africa's participation (outside of ancient Egypt) in the mathematical ferment that occurred in several eras of world history. Why?

Factors Underlying Mathematical Growth

Let us examine the factors that characterized the periods of great mathematical activity and of scientific inquiry in general. The first requirement is the existence of a well-organized society, with a level of agricultural development sufficient to produce a surplus that can support a division of labor. Along with metal workers, weavers and other craftsmen, there are merchants to exchange the surplus goods, as well as intellectuals with time to devote to scientific pursuits. The cultural interaction resulting from commercial contacts with other societies is a stimulus to scientific growth. The anthropologist, Leslie A. White, summarized these conditions: "Inventions and discoveries are much more likely to take place at culture centers, at places where there is a great deal of cultural interaction, than on the periphery, in remote or isolated regions" (page 2361).

New ideas may pose a threat to the Establishment, whether it be the guardians of the ancestral spirits, the divine king, the established church, or profiteers from the existing system. The innovators must be bold in counteracting restrictions placed upon them by the defenders of traditional thought.

One of the earliest spurts of mathematical development that we know of took place in Mesopotamia, the crossroads of the ancient world, a center of unrestricted commercial activity. In a later era, the ancient Greeks, living in an atmosphere of freedom from political and religious despotism, were perhaps the first to evolve a logical mathematical system.

A thousand and more years ago the Islamic world was the scene of mathematical growth. The Arabs contributed original work, as well as translating the neglected manuscripts of the Greek mathematicians into Arabic, thus preserving them until they were reintroduced into Europe centuries later in their Arabic versions. The Arabs have been called "cultural middlemen"; they traveled throughout Asia, north and east Africa and the Sudanic region, and southern Europe, spreading the contributions of Indians, Chinese, Persians, Jews, Arabs and Europeans to every region.

Islamic science developed against a background of wide commercial contacts, flourishing cities, and official encouragement of its growth. In Europe, on the other hand, medieval Christendom was rooted in reverence for the traditional, and in reluctance to challenge the authority of the Catholic Church. In 1202 Leonardo Fibonacci, on his return from his travels in the East, introduced Hindu numerals into Western Europe, but it took three centuries for the system to be widely adopted. Merchants in the growing cities of Italy were the first to profit by this method of writing numbers. Now the rules of computation, hitherto a mystery to all except the most learned, were comprehensible to the average keeper of commercial accounts. At this time European universities were offering not much more than some arithmetic and a bit of Euclidean geometry, translated into Latin from Arabic.

Not only was there little progress in mathematics in Europe outside of the commercial centers, but the subject was viewed with general suspicion by the Establishment. Among the crimes for which the Spanish Inquisition inflicted life imprisonment or death were the possession of Arabic manuscripts and the study of mathematics. "Mathematics was looked upon with fear because of the magical power of numbers." As late as the seventeenth century, the time of the persecution of Galileo, "mathematicians were denounced as the greatest of all heretics" (Homer W. Smith, page 261).

It was the growth of the industrial class in most of Europe, with its challenge to the existing order, which sparked the greatest age of mathematical development.

Islamic Culture Centers in Africa

Because of the diversity of African cultures, it is impossible to discuss Africa as a whole in relation to the question of scientific growth. For example, living side by side in Kenya today are two ethnic groups having entirely opposite attitudes toward cultural change. A few years ago the Kikuyu fought bitterly for participation in the industrialization of their land and for formal European-style schooling for their children. The Maasai, on the other hand, have remained aloof from European influence; even now many Maasai boys spend their days herding cattle, rather than attending school. Another instance—the Igbo of Nigeria have long been known as innovators, in contrast to the many tradition-bound African societies. Some areas of the continent have come under the influence of Islamic thought, others have had contact with Christian missionaries, and still others have scarcely been affected by foreign cultures.

Were there in Africa the kind of culture centers in which one might expect to find inventions and discoveries? Timbuktu, in ancient Mali and, later, part of Songhai, was a center of commerce and culture, an international city. The Islamic mosque of Sankore, built in the early fourteenth century, was virtually a university, and attracted scholars from the Mediterranean coastal regions and even from Asia. During the fifteenth and sixteenth centuries, the curriculum included law, rhetoric, surgery, astronomy, history and geography. The North African traveler and historian, Leo Africanus, wrote of Timbuktu in the early sixteenth century: "Here are a great store of doctors, judges, priests and other learned men, that are bountifully maintained at the king's cost and charges. And hither are brought divers manuscripts of written books out of Barbary, which are sold for more money than any other merchandise." The cultural level is indicated in the seventeenth century book *Tarikh es-Soudan*, the history of the Sudan, which quotes from a biography (written by his grandson) of al-Hajj Ahmed, a fifteenth century "lawyer, lexicographer, grammarian, prosodist, and scholar. He occupied himself with the sciences all his life. . . . At his death he left about seven hundred volumes [of his own writing]" (Hodgkin, page 91).

The medieval Sudanese states maintained a corps of clerks who wrote and copied manuscripts, supplementing the original writings of scholars. A few manuscripts, such as the *Tarikh es-Soudan*, have been translated into European languages. Hundreds more, having survived both the disorganization of western Sudanic society following the Moroccan invasion in 1591 and the attacks of termites, lie in libraries and homes, awaiting translation.

But as for mathematics, it was "somewhat neglected" at Timbuktu (Hogben and Kirk-Greene, page 81). Islamic learning of that period and later was dominated by theology and tradition; the sciences to which scholars like al-Hajj Ahmed devoted their lives did not lead to new mathematical developments, as far as we know.

By 1700 Katsina, one of the Hausa states in northern Nigeria, had succeeded Timbuktu as a center of learning. This city was the home of Muhammad ibn Muhammad, teacher of astrology, numerology and the number theory associated with these mystic arts, and author of the work on magic squares discussed earlier in this book. His teacher, Muhammad

Alwali of Bagirmi (near Lake Chad), wrote verses concerning divination, magic and astrology, none of which have survived.

Early in the nineteenth century the Fulani, under the leadership of Uthman dan Fodio, conquered large regions of northern Nigeria, bringing economic and political unity to these areas and extending the influence of Islam. With them came a new stimulus to learning and literature, for the leaders of the *jihad*, or holy war, were Muslim scholars first, and politicians or generals second. They imposed Arabic as the official language, and encouraged education in the traditional Islamic sciences. Their leader, Uthman dan Fodio, himself taught his younger brother arithmetic, among other subjects, according to the journal written by the latter.

Sultan Muhammad Bello, the son of Uthman dan Fodio, was noted as a man of letters. In 1826 the English traveler, Hugh Clapperton, paid a visit to the Sultan, bearing with him numerous gifts from the King of England, including a copy of Euclid's *Elements* in Arabic. Subsequently, he wrote in his journal (page 198): "Saw the Sultan this morning, who was sitting in the inner apartment of his house, with the Arabic copy of Euclid before him, which I had given to him as a present. He said that his family had a copy of Euclid brought by one of their relations, who had procured it in Mecca; that it was destroyed when part of his house was burnt down last year; and he observed, that he could not but feel very much obliged to the king of England for sending him so valuable a present. . . ."

Thus far, we have found no evidence of original work in pure mathematics in the African centers of Islamic learning. Possibly Muslim Africans participated in the mathematical activities of the Islamic cities in Spain, Egypt and Asia. A question to this effect, directed to a prominent scholar of the history of Arabic mathematics, evoked the reply that he could not answer it, since Islam makes no distinctions of race. All believers in Islam are required to read the Qur'an (Koran) in the Arabic language, and it is perfectly natural that they have Arabic names and write in that tongue, just as Europeans used Latin several centuries ago; the famous mapmaker Mercator was a Flemish subject named Kaufman—"merchant."

Obstacles to the Growth of African Mathematics

We know nothing of the contributions of non-Muslim Africans to the development of pure mathematics. Much of Africa's past has come to light in just the last few years, and no doubt much more will be discovered. But archaeological research has barely begun in most parts of Africa, surviving documents lie untranslated or unpublished, and in many cultures oral tradition is the only history. Some of the most important archaeological sites were destroyed by Europeans in their search for loot.

Nevertheless, some writers have concluded on the basis of available records that in the fifteenth century, the level of culture among the masses of black people in West Africa was higher than that of northern Europe during the same period. Why, then, do we not see in Africa the spurt in mathematical development that took place in Europe during the last five centuries? The reasons can be summarized as in the first place, geographical problems, and, secondly the destruction of African societies by invasion, the slave trade, and the scramble for Africa.

The geography of much of Africa—high mountains, arid desert, rain forest, unnavigable rivers, few natural harbors—makes it an inhospitable continent, discouraging the growth of stable agricultural and urban communities and extensive world trade. The malaria mosquito and the tsetse fly decimated the populations of man and beast. The soil of some areas is useful for building homes, but poor for the growth of crops.

Since the continent was sparsely populated, whole tribes often migrated long distances to find better land. Frequently, when people moved away, the wilderness came in behind them and raised its barriers, forcing these tribes to evolve their own way of life in relative isolation from the outside world. Much of the continent, notably eastern and southern Africa, has had stable communities for only the past few centuries.

For many societies, the economy was one of little more than bare subsistence. Of course there were exceptions, as in West Africa, but even where empires flourished, the way of life of people in the hinterland was scarcely affected by the wealth and display of the governing classes. Many people lived in a "carefully egalitarian world where personal gain above the level of the accepted norm would be a source of unhappiness or danger, since exceptional achievement could be only at the expense of one's neighbors" (Davidson, 1969, page 66). Individuals in such a well-ordered society do not seek wealth or social superiority that might create too great a departure from traditional tribal life. If cultural interaction is a requirement for invention, we can hardly expect to find scientific and mathematical inquiry in these isolated and tradition-oriented societies.

Added to the physical problems have been the numerous invasions of the continent. From the eleventh century onward, West Africa was the scene of Muslim holy wars and wars for wealth. In the fifteenth century the European slave trade in West Africa was already sizeable, but it was not to reach its peak until three centuries later, when the demands of the New World were greatest. A conservative estimate gives twenty-four million as the number of slaves exported from West Africa and Angola. Dr. W. E. B. Du Bois believed that one hundred million Africans were lost in the course of the slave trade. Many revolted, died in battle, jumped overboard, or perished in the infamous "Middle Passage," the journey to the New World, where the institution of slavery became the foundation for American prosperity.

The destructive effects of the slave trade upon African social organization can hardly be described adequately. Furthermore, it brought about a complete reversal of attitude on the part of Europeans towards Africans. Black people were read out of the human race, their culture and history eradicated from European thought. Africa became the "Dark Continent."

To complete the destruction, in the nineteenth century the scramble for Africa began. The European powers turned to Africa for new markets and colonies, and by 1900 they had completely redrawn the African map, with no regard for existing boundaries, societies, or political organizations. Colonial officers were sent in to set up governmental structures most acceptable to the European rulers; industrialists used African labor for greatest possible profit to themselves.

Further disruption came with the missionaries, who demanded that Africans renounce their traditional beliefs. Kenyatta proclaimed angrily: "Along with his land they rob him of his government, condemn his religious ideas, and ignore his fundamental conceptions of justice and morals, all in the name of civilization and progress" (page 318).

The missionaries introduced European-oriented education in their schools, where Africans "were taught to be loyal to the Government as a Christian virtue and to accept and respect their masters" (Mchunguzi, page 24). There was little continuity between the school and the African child's social environment. Since the subject matter had little relevance to the child's life, he was not motivated to learn. This accounted for the stress on rote memory which characterized the education of African children.

The irrelevance of mathematics education is exemplified in an incident related by Gay and Cole. The schools of which they write are conducted in English, and the child must spend his first year at school merely learning the language. Outdated American books and methods are used. For example, the child is required to memorize the multiplication tables, but does not learn to apply this information to problems he encounters in everyday life. One student, called upon to recite his tables, began, "Dah-de-dah-de-dah, dah-de-dah-de-dah." The teacher interrupted to ask what he was saying. The child replied that he knew the tune, but had not yet learned the words!

In the Belgian colonies no more than the minimum education was permitted. At the time of the Congo's independence, in June 1960, there were fewer than twenty college graduates in the whole country, while in that same year just 152 Congolese completed secondary school (see Fullerton).

Africa's Contribution to World Culture

Western researchers are struck by the contrast in moral outlook between African cultures and those of the West. In his report on "Psychology,"

A. C. Mundy-Castle writes (page 252): "[African societies] promote social integration, friendliness, respect for others, serenity, social ease, and a capacity for pleasure. . . . In contrast the supertechnological societies thrive on individualism, assertiveness and aggressive competition." Unbridled technological growth uninhibited by moral strictures is characteristic of our society. Basil Davidson comments bitterly (1969, page 115): "Science predicts disaster with the continued spread of nuclear weapons, but the spread continues . . . because the mandatory moral force to stop it is no longer there." Mundy-Castle expresses the hope that the rest of the world learns from Africa as Africa acquires the achievements of the "supertechnological" societies. "The wider world may benefit from its integration with Africa, by constructive modulation of aggressive and receptive tendencies, giving rise to a nobler and more harmonious expression of human intelligence" (page 253).

But African social patterns have their own characteristic dangers. In the strictly-ordered African society, "conformity to prescribed behavior became the only way of doing what was 'right and natural'" (Davidson, 1969, page 115). Although the ideology of the African society was often "successful in achieving social harmony, a heavy price was paid in conservative conformism."

On the eve of the Second World War the black poet Aimé Césaire, appalled by the consequences of Europe's progress, cried out:

Hurrah for those who have never invented anything
for those who have never explored anything
for those who have never conquered anything . . .

Africans today are inventing and exploring. Perhaps they will succeed in combining technological progress with their traditions of morality and communal concern—if Africa accomplishes this, it will have contributed not only to technological advancement, but, more importantly, to the sphere of human relations.

APPENDIX

ACKNOWLEDGMENTS

It is a great pleasure to acknowledge the assistance I have received in the preparation of this book. I should like to express my gratitude to the many people who took valuable time to reply to my inquiries with helpful information: J. F. A. Ajayi, J. A. Akinpelu, Samuel Ayany, J. A. Ayorinde, J. A. Ballard, A. D. H. Bivar, Frank E. Chapman Jr., George Crossette, Basil Davidson, Jean de Heinzelin, Alison Des Forges, A. A. Dike, Michele Eberhart, Harry Epstein, Ronald Frankenberg, John Gay, Gordon D. Gibson, Joseph H. Greenberg, Lawrence W. Henderson, Sidney J. Hogben, Richard Hull, Cecilia Irina, Justin Irina, Graham Irwin, Marion Johnson, Ruth Jones and the staff of the International African Institute, Ernest Kaiser and the staff of the Schomburg Collection of the New York Public Library, Ole Kantai, Jim Karnes, E. S. Kennedy, Shadrack Kimalel, B.E. Kipkorir, Alexander Marshack, John Mbiti, James Mhina, Onesmo ole Moiyoi, Philip Morrison, David Mulumba, Kivuto Ndeti, J. H. Nketia, Offia Nwali, H. M. A. Obayemi, R. G. Opondo, D. N. Oransaye, Richard Pankhurst, Labelle Prussin, O. F. Raum, Paul Rosenblum, Jan Rubinowitz, Marjorie Spiers, Dirk J. Struik, David Zarembka and Rodah Zarembka. I am indebted to Ruth Soika, Nancy James, Gerda Lederer and Selma Teich for assistance with translations, and to Tillie Godt for loving help with typing. Edward Telesford of the British Museum, Barbara Stuckenrath of the Smithsonian Institution, and my friend Gerald Ames were invaluable for their assistance in the selection and preparation of photographs.

I appreciate the interest shown by John Henrik Clarke in reviewing the manuscript and writing the introduction. Had it not been for the encouragement of Howard Eves, this book would not have come into existence. Most of all I am grateful to D. W. Crowe for a wealth of information and for suggestions at every stage of progress, as well as for his reading of the completed manuscript; to my friend Rose Wyler for her unfailing interest in this project and her reading of the manuscript, and to the men in my family: my son Thomas for painstaking editing, my son Alan for information and African contacts, and my husband Sam for photographs, for helping in every possible way, and for his devotion and patience during this period.

I am grateful to the following for permission to reproduce copyright material:

Extracts from Aimé Césaire's *Cahier d'un Retour au Pays Natal* published by Présence Africaine, Paris (1956). The translation of these lines appears in *The Horizon History of Africa*, published by American Heritage Publishing Co., Inc. (1971).

Excerpts from "Science and Africa" by Frank E. Chapman, Jr. are reprinted by permission of *Freedomways* magazine, Vol. 6, No. 3 (1966), 799 Broadway, New York City, N.Y.

Item 9° from *In Mathematical Circles*, by permission of the author Howard Eves, and the publisher Prindle, Weber & Schmidt, Inc.

Excerpts from *Fourteen Hundred Cowries*, collected by Abayemi Fuja and published by Oxford University Press.

Extracts from *The History of the Yorubas: From the Earliest Times to the Beginning of the British Protectorate* by Rev. Samuel Johnson (1921) by permission of the publisher, Routledge & Kegan Paul Ltd.

Excerpts from *African Religions and Philosophy*, by Rev. John S. Mbiti, by permission of the publishers Frederick A. Praeger, Inc. and Heinemann Educational Books Ltd.

Translated selections and map from Marianne Schmidl's "Zahl und Zählen in Afrika" by permission of the Anthropologische Gesellschaft in Wien, Austria.

Information about the game Oh-Wah-Ree by permission of 3M Company, St. Paul, Minnesota.

Two stanzas of "The Round Mud Hut" by Joseph Waiguru, in *Origin East Africa*: ed. by David Cook, by permission of the publisher, Heinemann Educational Books Ltd.

Selections from "The Locus of Mathematical Reality: An Anthropological Footnote" by Leslie A. White, in *The World of Mathematics*: ed. by James R. Newman, by permission of the publisher, Simon & Schuster, Inc.

Excerpts from "Black African Traditional Mathematics" by Claudia Zaslavsky, reprinted from the *Mathematics Teacher*, April 1970 (vol. 63, pp. 345–56), ©1970 by the National Council of Teachers of Mathematics. Used by permission.

Frank E. Chapman, Jr. has given permission to use selections from his unpublished manuscript and his autobiography.

Chief J. A. Ayorinde has kindly consented to quotation from his unpublished work.

The author and publishers make grateful acknowledgment to the following for photographs and drawings:

David Attenborough for Figure 13–16.

The British Museum for Figures 1–1; 2–5, 2–6; 7–3, 7–4; 8–3, 8–4; 10–6; 11–1, 11–3, 11–4, 11–5, 11–7; 14–1, 14–6, 14–7, 14–8, 14–10, 14–14, 14–15, 14–18, 14–19, 14–20; 15–1; 17–1, 17–2; 19–1.

Creative Playthings, Princeton, N.J., for Figure 11–10.

D. W. Crowe for Figures 10–1; 14–21, 14–22, 14–23, 14–24, 14–25, 14–26.

J. de Heinzelin for Figure 2–1.

International African Institute for the line drawing in Figure 14–2, from *Conversations with Ogotemmêli* by M. Griaule.

The Museum of Primitive Art, New York, for Figures 14–5 and 14–17.

National Development Corporation, Tanzania, for Figure 11–9.

National Museum of Tanzania, Dar es Salaam, for Figures 1–5; 7–5, 7–6; 9–1; 20–1.

The Peabody Museum, Harvard University, for Figure 1–4.

Labelle Prussin for Figures 13–14 and 13–17.

O. F. Raum for Figure 3–1B.

Smithsonian Institution, Washington, D.C., for Figures 6–1; 7–1; 14–9; 14–11.

All other photographs were taken by Sam Zaslavsky.

CHAPTER REFERENCES AND NOTES

The Notes include both parenthetical remarks and information which came to the author's attention as this book was going into production.

SECTION 1

Chapter 1   AFRICAN MATHEMATICS?

Achebe, 1967.
Cajori, 1896, 1917.
Chapman,
   in *Freedomways*, Vol. 6,#3.
Conant, 1896.
Dantzig, 1940.
Hogben, 1960.
Hollis, 1905.
Kantai, Ole,
   personal communication.
Lévy-Bruhl, 1910, 1966.
Mann, 1887.
Menninger, 1958, 1969.
Moiyoi, Onesmo ole,
   personal communication.
Ngugi, 1965.
Schmidl, 1915.
Seidenberg, 1960.
Smeltzer, 1958,
   does mention some African numerals.
Smith, D. E., 1923, 1958.
White, 1956.
Wilder, 1968.

NOTE on the scientific method.
   After having examined the known data, the scientist makes a tentative guess, or hypothesis, that he believes can explain his observations. The next step, and perhaps the most crucial one, is the objective testing of the scientist's hypothesis. Many an able investigator has spent long years in trying to prove an incorrect hypothesis. However, his efforts are not wasted; they contribute to the greater body of information available to other workers in his field.
   Only when the scientist is sure that his hypothesis is correct and that some equally plausible hypothesis could not explain the same data, is he justified

in calling it a scientific law. This new scientific principle is open to the most careful scrutiny on the part of his fellow workers, leading to further hypotheses and testing.

Chapter 2 HISTORICAL BACKGROUND

Ben-Jochannan, 1970.
Bernal, 1954.
Bohannan, 1964.
Boyer, 1968.
Clark, J. Desmond,
  "Archaeological Studies", in Brokensha and Crowder, 1967.
Collins, Robert O.,
  "Problem V. African States", in Collins, 1968.
Davidson, 1969.
Davidson, 1966.
DeGraft-Johnson, 1954.
De Heinzelin, 1962.
Dike, K. O., 1956.
Diop, Cheikh Anta,
  "Negro Nations and Culture," in Collins.
Du Bois, 1956.
Eves, *History*, 1969.
Greenberg, 1966.
  He states that the term 'Hamitic' does not refer to any valid linguistic entity, and criticizes the use of the term by earlier Africanists. "The vagueness of the use of the term Hamite as a linguistic term and its extension as a racial term for a type viewed primarily as Caucasoid, has led to a racial theory in which the majority of the native population of Negro Africa is considered to be the result of a mixture between Hamites and Negroes. With this is often combined a belief either in the inherent superiority of the Hamite element or in a factual estimate that it has shown itself everywhere as a conquering, predominantly pastoral element among Negroid agricultural peoples" (page 49). Greenberg does not use the term 'Hamite' to describe any linguistic family, because of the many false implications of the word.
Hogben, L., 1960.
James, 1954.
Marshack, Alexander,
  personal communication.
  ———, 1972.
Murdock, 1959.
Neugebauer, 1962.
New York Times
  "Forty-three Thousand-Year-Old Mine Discovered in Swaziland." February 8, 1970.

Ogot and Kieran, 1969.
Phillipson, 1970.
Reinhold, Robert.
  "Bone Traces Man Back Five Million Years." *New York Times*, February 18, 1971.
Shinnie, 1965.
Struik, 1967.
Van der Waerden, 1961.
Willett, 1971.

NOTE on the wealth of southern Africa.

Long before the tenth century, gold and ivory from the present region of Zimbabwe (Rhodesia) were brought to the port of Sofala and exchanged for Indian beads. Arab, and subsequently Persian, merchants distributed the precious goods to India and other lands of the East. In the sixteenth century the Portuguese sought to take over this valuable trade.

The Portuguese writer Joao de Barros gave the following description of the Emperor of Monomotapa's palace:

> The inside consists of a great variety of sumptuous apartments, tapestry, the manufacture of the country. The floors, ceiling, beams and rafters are all either gilt or plated with gold and curiously wrought, as are all the chairs of State, tables, benches, etc. The candlesticks and branches are made of ivory inlaid with gold, and hang from the ceiling by chains of the same metal or silver gilt ... In short, so rich and magnificent is this palace, that it may be said to vie with that which distinguishes a monarch of the East.

SECTION 2

Chapter 3   CONSTRUCTION OF NUMERATION SYSTEMS

Bernal, 1954.
Bosman, 1704, 1967.
Chapman,
  unpublished manuscript.
Clapperton, H.,
  in Hodgkin, 1960.
Crowder, 1968.
Delafosse, 1928.
Einzig, 1966.
Eves,
  In Mathematical Circles, 1969.
Gay and Cole, 1967.
Hogben, L., 1960.

Innes, 1967.
Irina, Cecilia,
    personal communication.
Jacottet, 1927.
Kramer, 1951,
    mentions the Sotho expression for ninety-nine.
Menninger, 1958, 1969.
Migeod, 1911.
Neugebauer, 1962.
Turner, 1949, 1969.
Wilder, 1968.

NOTE on the Damara sheep-trading story.
In the time of Sir Francis Galton, who related this incident, British scientists maintained that human intelligence was directly correlated with the size of the brain, and therefore of the cranium. While urging the members of the Royal Anthropological Society to continue their investigations in the field of intelligence by comparing head sizes of members of "upper" and "lower" social classes, Galton made a statement to the effect that it is a well known fact that university students wear caps of larger head size than do working men (*Journal of the Royal Anthropological Society*, XVI (1887)).

NOTE on arithmetic skill of the Dahomean traders.
The Dutch trader William Bosman wrote at the beginning of the eighteenth century (page 352): "They are so accurately quick in their Merchandise Accompts, that they easily reckon as justly and as quick in their Heads alone, as we with the assistance of Pen and Ink, though the Summ amounts to several Thousands; which makes it very easie to Trade with them."

Chapter 4   HOW AFRICANS COUNT

Abraham, 1967.
Almeida, 1947,
    on numeration in Guinea-Bissau.
Armstrong, 1967.
————, 1962.
Delafosse, 1929.
————, 1928.
Dike, A. A.,
    personal communication. I am indebted to Mr. Dike for information about the Igbo people and their numeration.
Döhne, 1857.
Eberhart, Michele,
    personal communication about Hausa gestures.
Gay and Cole, 1967.
Greenberg,
    "Niger-Congo", in Collins, 1968.

Herskovits, 1939.
Howeidy, 1959.
Igwe and Green, 1967.
Innes, 1967.
Irina, Cecilia,
   personal communication about Sotho numeration.
Jacottet, 1927.
Leonard, A. G.,
   The Lower Niger and Its Tribes (London: 1906), quoted in Schmidl.
Mathews, 1964.
Menninger, 1958, 1969.
Migeod, 1911.
Raum, 1938.
Schmidl, 1915.
Seidenberg, 1960.
Smith, D. E., 1923, 1958.
Thomas, 1920.
Ziervogel and Louw, 1967.

NOTE on Mande-speaking peoples.
A large number of western Sudanic peoples belong to the Mande-speaking family, perhaps the oldest language family in West Africa. To this group belonged the founders of the ancient lands of Ghana and Mali. Included in the family are those people called Mandingo or Malinke, the Bambara, the Dogon, the Soninke, the Diula or Wangara, the Vai, the Mende and the Kpelle.

Chapter 5 TABOOS AND MYSTICISM

Achebe, 1967.
Bernal, 1954,
   for the anecdote about the philosopher Hegel.
Bleeker, 1969.
Davidson, 1969.
Delafosse, 1929.
Dike, A. A.,
   personal communication.
Epstein, Harry,
   personal communication about Hebrew taboos.
Frobenius, 1916, 1968.
Gay and Cole, 1967.
Griaule, 1965.
Henderson,
   personal communication.
Herskovits, 1952.
Kramer, 1955,
   about Hebrew taboos.

Mead, Margaret:
   "A Naturalist at Large: The Island Earth," *Natural History Magazine*, New York,
   LXXIX 1 (Jan. 1970).
Migeod, 1911.
Parrinder, 1967.
Rattray, Vol. II, 1927, 1954.
Schmidl, 1915.
Seidenberg, 1962.
———, 1960.
Wilder, 1968.
Williamson and Timitimi, 1970.

SECTION 3

Chapter 6   THE AFRICAN CONCEPT OF TIME

Achebe, 1967.
Bohannan, 1968.
Davidson, 1969.
Des Forges, Alison,
   personal communication.
Hill, 1966
Kenyatta, 1938, 1953.
Mbiti, 1969.
Murdock, 1959.
Nilsson, 1920.
Ojike, 1946.
Ukwu, 1967.

Chapter 7   NUMBERS AND MONEY

Achebe, 1959.
Akinpelu, J. A.:
   personal communication. He related the anecdote about the Yoruba who
   understated the number of people living in their community.
Barth, 1857, 1965.
Basden, 1921.
———, 1938.
Bohannan, 1968.
Bosman, 1704, 1967.
Clegg, 1969,
   He writes that Christopher Columbus referred in his journals to his discovery
   in America of black men from the Guinea Coast, who carried spears tipped

with "guanin," a gold alloy which turned out, when it was assayed, to be of the very same composition as that found in West Africa.

Davidson, 1969.

————, 1966.

————, 1964.

De-Graft Johnson, 1954.

Delafosse, 1929.

Des Forges, Alison,
  personal communication about Rwandese taxation.

Dike, A. A.,
  personal communication.

Du Bois, 1965.

Einzig, 1966.

Hall, B., 1964.

Herskovits, 1952.

————, 1962.

Ibn Battuta, 1929.

Jeffreys, 1938.

Johnson, M., 1970 (Parts I and II).

————,
  personal communication.

Kirk-Greene, 1960.

Lander, R. and L., in Hodgkin, 1960,
  on the Igbo woman's silent trade in yams.

Al-Maqrisi,
  "Kanem-Bornu at the Height of its Power" in Hodgkin, 1960.

Mbiti, 1969.

Murdock, 1959.

Oliver, R.,
  "The Problem of the Bantu Expansion" in Collins, 1968.

Rattray (I), 1923, 1969, gives a long list of the names of the goldweights and their equivalents in British currency, ranging from one-third of a penny to eighty pounds.

Raum, 1938.

Richards, Audrey I.,
  "The Ganda," in Collins, 1968.

Ryder, 1965.

Shinnie, 1965.

Zarembka, David,
  personal communication. I am indebted to him for information about the Muyenzi district, Tanzania.

NOTE: Schmidl has this to say on the subject of *keme*:
"Moreover, with the increasing influence of Islam and the concomitant spread of the decimal system, the same names came to be used for different quantities. Due to the use of strings of cowries, 80 (= *keme*), meaning a large quantity,

became the most common round number among the Malinke; the same word was used by the Mohammedan Mande for 100. The calculations became even more complicated, since the Soninke expressed 60 as *keme*. One differentiated between *keme* of Islam (= 100), of the Malinke (= 80), and of the Soninke (= 60). In connection with the last two, it should be noted that the next higher units were constructed on the basis $8 \times 80 = 640$ (by the Malinke) and $6 \times 60 = 360$ (by the Soninke)" (page 194).

NOTE on goldweights.
In 1701 Bosman gave a simplified outline of the goldweights and their European equivalents (pages 85–86):

24 Damba (*damma*) ⎫ = one *ackey*   = 1/16 ounce gold
12 Tacoe (*taku*)   ⎭   = 5 shillings (British)   (5 sh.)

    4 Angels(*ackey*) = one *peso*   = 1/4 ounce gold = one pound.   (£1)
    8 *peso*   = one *benda* = 2 ounces gold = 8 pounds.   (£8)
    4 Bendo (*benda*) = one mark   = 8 ounces gold = 32 pounds.   (£32)
    2 marks   =   = 16 ounces gold = 64 pounds.   (£64)

NOTE on copper rod currency.
According to Latham, this currency has been in use in Africa for many centuries, long before the penetration of Europeans. Besides Ibn Battuta's description of such a currency in the Niger region in the mid-fourteenth century, "the large number of copper wires unearthed at the Igbo-Ukwu excavations, dated to the tenth century, may also have formed a similar currency, particularly as no cowries or other articles which might have formed a currency were found." European travelers in Africa reported that rods were split into wires for the purchase of small items.

NOTE on the extension of credit.
Bosman commented about the slave trade in Dahomey (page 363a): "But if there happen to be no stock of Slaves, the Factor must then resolve to run the Risque of trusting the Inhabitants with Goods to the value of one or two hundred Slaves; which Commodities they send into the In-land Country, in order to buy with them Slaves at all Markets, and that sometimes two hundred Miles deep in the Country." In a letter dated 1702, and included in Bosman's *Description of the Coast of Guinea*, David van Nyendael wrote about Benin (pages 433–434): "Another Inconvenience, which really deserves Complaint, is, that at our Arrival here, we are obliged to trust them with Goods to make Panes or Cloaths of; for the Payment of which we frequently stay so long, that by reason of the Advancement of the Season, the Consumption of our Provisions, and the Sickness or Mortality of our Men, we are obliged to depart without our Money: But on the other hand, the next time we come hither, we are sure to be honestly paid the Whole." Trade at Benin was usually for elephants' tusks. The rulers of Benin refused to have anything to do with the slave trade at that time.

For a discussion of credit, see C. W. Newbury, "Credit in Early Nineteenth Century West African Trade," *Journal of African History,* Vol. XIII #1(1972), pages 81–95.

Chapter 8 THOSE FAMILIAR WEIGHTS AND MEASURES!

Brain (undated).
Davidson, 1969.
Davidson, 1961.
Einzig, 1966.
Farquhar, 1948.
Gay and Cole, 1967.
Idehen, Peter,
    unpublished manuscript.
Leiris and Delange, 1968,
    about Asante goldweights.
Pankhurst, 1969.
Rattray (I), 1923, 1969.
Raum, 1938.
Ritchie-Calder, 1970.
Roscoe, 1911.
Trimingham, J. S.,
    "West Sudan States," in Collins, 1968.

MEASURE
In a communication to the author, Dr. Crowe writes: "In Nigeria as recently as ten years ago (and perhaps still today) the cigarette tin was a standard unit of measure in the market. In *Miss Williams' Cookery Book* (R. O. Williams, Longmans, Green and Co., 1957), a book written by a Nigerian for use by Nigerians, many of the measurements in the recipes are given in terms of cigarette tins. For example, the recipe for *alapa* on page 84 begins:

> 1 cigarette tin beans
> 1/2 cigarette tin *egusi*
> Pepper
> Salt
> Water

Chapter 9 RECORD-KEEPING: STICKS AND STRINGS

Achebe, 1959.
Dantzig, 1940,
    for the Dickens story of the British tally sticks.
Farquhar, 1948.
Lagerkrantz, 1968–69.

Ndeti, K.,
   personal communication about the Kamba.
Olderogge, 1966.
Rattray (II), 1927, 1954.
Raum, 1963.
Ibid., 1963.
Roscoe, 1911.
Van Sicard, 1954.
NOTE on the census in Dahomey.
   For a fascinating and detailed description of the census-taking process, taxation,
   and other aspects of life in Dahomey, see Melville J. Herskovits: *Dahomey, an
   Ancient West African Kingdom* (2 Volumes). (Evanston, Illinois: Northwestern
   University Press, 1967.)

SECTION FOUR   INTRODUCTION

Des Forges, Alison,
   personal communication about the poetry of Rwanda.

Chapter 10   GAMES TO GROW ON

Basden, 1938.
Bastin, 1961.
Bell, 1960,
   on the Asante version of Tic-Tac-Toe.
Blacking, 1967.
Crowe, D. W.,
   for translation of Jokwe story.
Dike, A. A.,
   personal communication on Igbo gambling.
Farquhar, 1950.
Gay and Cole, 1967.
Matthews, 1964,
   on Zimbabwe stone games.
Mott-Smith, 1954.
Raum, 1938,
   on Tanzanian games.
Rohrbough, 1936.
Siegel, 1940.
Torday, 1925,
   on Bushongo (Shongo, Kuba) networks.

Turnbull, 1961.
Williamson, John, 1943.

Chapter 11   THE GAME PLAYED BY KINGS AND COWHERDS—
AND PRESIDENTS, TOO!

Adamson, 1961.
Ayany, Samuel,
   personal communication on Luo game.
Ayorinde, J. A.,
   personal communication on Yoruba game.
Basden, 1921.
Bell, 1960.
Crowe, D. W.,
   Entebbe Mathematics Workshop material.
Culin, 1894.
Delano, 1937.
Des Forges, Alison,
   personal communication on Rwandese game.
Dike, A. A.,
   personal communication on Igbo game.
Driberg, 1927.
Haggerty, John B. "Kalah." *Arithmetic Teacher* 11 (May 1964): 326–330.
Hall, R. de Z., 1953.
Herskovits, 1932.
Jensen, 1936,
   on Ethiopian phallic megalith.
Lanning, 1956.
Lasebikan, 1963.
Listowel, Judith:
   Making of Tanganyika (London: Chatto and Windus, 1968), on President
   Nyerere's skill in *soro*.
Luschan, 1968.
Matthews, 1964.
Murray, 1952.
Nsimbi, 1968.
Ojike, 1946.
Rattray, Vol. II, 1927, 1954.
Rohrbough, 1955.
Rubinowitz, Jan,
   personal communication.
Sawyer, 1949.
Shinnie, 1965,
   on King Shamba Bolongongo of the Kuba people.

TIME Magazine:
  "Pits and Pebbles," Vol. 81 #24 (June 14, 1963).
Torday, 1925.
Torrey, 1963.
Wayland, 1963.
Williamson and Timiti, 1970.

NOTE on Turkana children.
  In a footnote to his article, "Archeology in the Turkana District, Kenya" (*Science*, 176, #4033, 28 April 1972), Lawrence H. Robbins remarks: "Stone-Age artifacts are commonly collected by Turkana children, who are fond of playing games with brightly colored stones. As a result, small circles of chert and obsidian artifacts resembling features of archeological significance are frequently encountered on the surface. The children use these stones to represent imaginary domestic stock, or for an *omweso*-type game."

NOTE on the game in Dahomey.
  In his very informative book, *Dahomey, an Ancient West African Kingdom*, Herskovits writes (page 287): "When the parents of the girl discover that her first period has come, she is seated on a mat of native make . . . and an *adjito*— a game-board on which *adji* is played—is put before her. Friends come and play with her, and all who visit her as she sits there—that is, for the duration of her first period—must bring her a present of at least a few *sous*."

NOTE on the strategies used by good players among the K pelle of Liberia.
  "The winning player makes sure he has solid defenses, that he catalogues the possibilities of every move, that he reserves time to himself, that he lures the opponent into making premature captures, that he moves for decisive rather than piecemeal victories, and that he is flexible in redistributing his forces in preparation for new assaults." (*The Cultural Context of Learning and Thinking*, by Michael Cole, John Gay, Joseph A. Glick, and Donald W. Sharp. New York: Basic Books, 1971, page 184.)

Chapter 12   MAGIC SQUARES

Andrews, 1960.
Ball, 1928.
Bivar and Hiskett, 1962.
Bivar, A. D. H.,
  personal communication.
Cajori, 1929.
Crowe, D. W.,
  personal communication.
Grossman and Magnus,
  Groups and their Graphs. New York: Random House, 1964.
Gwarzo, 1967.
Hogben, S.,

personal communication.
Husseini, S.,
   to whom I am indebted for the translation of Muhammad Ibn Muhammad's manuscript.
Ibn Muhammad, Muhammad, 1793.
Struik, 1963.

SECTION 5

Chapter 13   GEOMETRIC FORM IN ARCHITECTURE

Davidson, 1966.
Eberhart, Michele,
   personal communication about the Fulani.
Fernandez, James W.,
   "Zulu Zionism", *Natural History*, LXXX #6 (June-July, 1971).
Fitch and Branch, 1960.
Frobenius, 1913, 1968.
Herskovits, 1962.
Hintze, 1965.
Leuzinger, 1960.
Lindblom, 1920.
Murdock, 1959.
Ngugi, 1965.
Prussin, 1969.
Prussin, Labelle,
   "Sudanese Architecture and the Manding." *African Arts*, U.C.L.A., III #4 (Summer, 1970).

———,
   personal communication.
Raum, 1938.
Roscoe, 1915,
   on the Lake Kyoga dwellers.
Routledge, 1910, 1968.
Shinnie, 1965,
   on ancient Meroë.
Turnbull, 1961.
Waiguru, Joseph,
   "The Round Mud Hut" in *Origin East Africa*, edited by David Cook (London: Heineman, 1965).
Willett, 1971.
Zarembka, Rodah,
   personal communication.

NOTE on Mosques.
Mrs. Prussin states that Islamic law prescribed the square form for the mosque. Although some of the Guinean mosques appear completely round on the outside, they are, in fact, square on the inside.

Chapter 14 GEOMETRIC FORM AND PATTERN IN ART

Crowe, 1971
Crowe, D. W.,
   personal communication
Griaule, M., 1965.
Himmelheber, Hans,
   "The Concave Face in African Art," *African Arts*, U.C.L.A., IV #3 (Spring, 1971).
Leiris and Delange, 1968.
Parrinder, 1967.
Trowell, 1960.
Willett, 1971.

NOTE on Kuba art.
Even today the concepts of form and pattern permeate the lives of the Kuba people, and their artists are very sophisticated in their ability to discuss their work from every point of view.

NOTE on textile patterns.
Many of the African patterns seen today in modern Western textiles are copied directly from anthropology books by the designers and manufacturers.

SECTION SIX

Chapter 15 HISTORY OF THE YORUBA STATES AND BENIN

Aderibigbe, A. A. B.,
   "Peoples of Southern Nigeria," in Ajayi and Esprie, 1968.
Apple, R. W., Jr.:
   "Singsong Voices, Bright Caps Mark Ibadan's Yorubas," *The New York Times*, June 14, 1969.
Babayemi, 1968.
Crowder, 1966.
Davidson, 1969.
Davidson, 1966.
Hodgkin, 1960.

Johnson, Samuel, 1921, 1966.
Lloyd, Awe and Mabogunje, 1967.
Shaw, 1967.
Shinnie, 1965.
Willett, 1971.

Chapter 16   SYSTEMS OF NUMERATION

Ajayi, J. F. Ade,
    personal communication.
Akinpelu, J. A.,
    personal communication.
Armstrong, 1962.
Ayorinde, J. A.,
    personal communication.
Conant, 1896.
Idehen, Peter,
    unpublished paper.
Johnson, Samuel, 1921, 1966.
Mann, 1887.
Munro, 1967.
Schmidl, 1915.
Thomas, 1910.

Chapter 17   SIGNIFICANT NUMBERS

Armstrong, Olayemi and Adu, 1969.
Bascom, 1969.
Crowder, 1966.
Dennett, 1910, 1968.
Egharevba, 1960.
Fagg and Willett, 1960.
Forde, 1951.
Fuja, 1962.
Idehen, Peter,
    unpublished paper.
Idowu, 1962.
Johnson, S., 1921, 1966.
Leuzinger, 1960.
Morton-Williams, 1964.
Ojo, 1966.
Stevens, 1966.

Chapter 18   TIME-RECKONING

Achebe, 1967.
Achebe, 1959.
Ajayi, J. F. Ade,
    personal communication.
Akinpelu, J. A.,
    personal communication.
Frobenius, 1913, 1968.
Hill, 1966.
Idehen, Peter,
    unpublished paper.
Ojo, 1966.

NOTE.

The Yoruba concept of the universe is expressed in the saying: "A mighty calabash of two hemispheres, never unlidded: the earth and the sky." The earth is
fixed on the lower hemisphere, and the sun, moon and stars move about on the
upper. Venus is the only heavenly body, besides the sun and moon, for which
the Yoruba have a name. There is a saying: "It is only the unobservant who concludes that Venus is the moon's dog just because they are sometimes found close
together; in fact, Venus is not the moon's dog."

Chapter 19   MARKETS AND CURRENCY

Basden, 1921.
Bohannan, 1964.
Bosman, 1704, 1967.
Conant, 1896.
Davidson, 1961.
Delano, 1963.
Dike, 1956.
Einzig, 1966.
Herskovits, 1952.
Hill, 1966.
Hodgkin, 1960.
Johnson, M., 1970 I and II.
Johnson, S., 1921, 1966.
Lloyd, P. C.,
    "Osifekunde of Ijebu" in Africa Remembered, Philip Curtin, Editor. (Madison,
    Wisc.: University of Wisconsin Press, 1967.)
Ojo, 1966.
TIME
    "African Women; From Old Magic to New Power," August 31, 1970.

SECTION 7

Chapter 20   HISTORY OF EAST AFRICA

Ayany, Samuel,
personal communication. I am indebted to Mr. Ayany for information about
the Luo.
Collins, Robert O.,
"African States" in *Problems in African History*, Robert O. Collins, ed., 1968.
Davidson, 1969.
Du Bois, 1965,
on the East African trade in slaves and ivory.
Freeman-Grenville, G. S. P.,
"The Kilwa State," in Collins, 1968.
Greenberg, J. H.,
"The Languages of Africa," in Collins, 1968.
Kimalel, Shadrack,
personal communication about the Nandi, a branch of the Kalenjin peoples.
On a personal note—Mr. Kimalel's father had been criticized by his neighbors
for sending his son to school when the boy rightfully should have been tend-
ing the herds. Years later, when he returned to his homeland as a teacher, and
subsequently as a school inspector, the neighbors agreed that the father had
indeed been wise.
Kipkorir, B.E.,
personal communication. Dr. Kipkorir was extremely helpful with regard to
information about the Kalenjin.
Ogot, 1967.
Ogot and Kieran, 1968.
Osogo, 1966.

NOTE.

The rite of circumcision as a part of the initiation into adulthood was acquired
through contact with the Cushitic people, probably from Ethiopia. This custom
was practised by the ancient Egyptians, and for thousands of years the Jews
have circumcised their sons on the eighth day after birth. The custom of circumcision and clitoridectomy was adopted by many Bantu and Nilotic peoples in
Kenya and northern Tanzania. In addition, the Maasai and the Kalenjin remove
two lower teeth. Among the Luo the only initiation rite is the removal of six
lower teeth.

Chapter 21   SPOKEN AND GESTURE COUNTING

Atkins, 1961.
Ayany, Samuel,
personal communication about the Luo.

Brain.

Gulliver, 1958.

Hattersley, 1908, 1968.

Hurel, 1951.

Irina, Justin,
   to whom I am indebted for information about the Taita.

Kantai, Ole,
   I am grateful to Mr. Kantai for material about the Maasai people.

Kimalel, Shadrack,
   personal communication about the Kalenjin.

Kipkorir, B.E.
   personal communication about the Kalenjin.

Mbiti, 1959.

Menninger, 1958, 1969.

Mulumba, David,
   I am indebted to Mr. Mulumba for material about the Kamba.

A. V. O., 1936.

Raum, 1938.

Roscoe, 1911.

Roscoe, 1915.

Schmidl, 1915.

Tucker and Mpaayei, 1955.

Williamson, 1943.

Zarembka, David,
   personal communication.

Zaslavsky, Alan,
   did very valuable field work while teaching secondary school in Kenya.

Chapter 22   NUMBER SUPERSTITIONS

Ayany, Samuel,
   personal communication about the Luo.

Gecaga and Kirkaldy, 1953.

Gulliver, 1958.

Hollis, 1909.

Irina, Justin,
   personal communication.

Kantai, Ole,
   personal communication about the Maasai.

Kenyatta, 1938, 1953.

Kimalel, S.,
   personal communication about the Nandi.

Kipkorir, B.E.,
   personal communication about the Kalenjin.

Lindblom, 1920.

Mbiti, John,
  personal communication.
Raum, 1938.
Roscoe, 1911.
Routledge, 1910, 1968.
Spiers, Marjorie,
  personal communication
Williamson, John, 1943.
Zaslavsky, Alan,
  interviewed East African secondary school students.

NOTE ON SEVEN.
A comparison of the number words from one to ten in several Bantu languages shows that the words are almost identical except for the numeral for seven. Here we find:

Kikuyu:	*mugwanja*
Kamba:	*muonza*
Taita:	*mufungade*
Rwanda:	*rindwi*

Chapter 23   EAST AFRICAN TIME

Ayany, S.,
  personal communication about the Luo.
Bleeker, 1964.
Brain (undated).
Davidson, 1969.
Dundas, 1926.
Dyson-Hudson, 1966.
Gulliver, 1963.
Herskovits, 1962.
Kenyatta, 1938, 1953.
Kipkorir, B.,
  personal communication.
Lindblom, 1920.
Mbiti, 1969.
Murdock, 1959.
Nilsson, 1920.
Ogot and Kieran, 1969.
Osogo, 1966.
Roscoe, 1915.
Routledge, 1910, 1963.
Zarembka, Rodah,
  personal communication.

Chapter 24 WEALTH MEANS CATTLE—AND OTHER LIVESTOCK

Davidson, 1969.
Des Forges, Alison,
  personal communication about currency in Rwanda.
Einzig, 1966.
Gulliver, 1963.
Herskovits, 1962.
Kenyatta, 1938, 1953.
Kipkorir, B.,
  personal communication about Kalenjin bridewealth.
Lindblom, 1920.
Mhina, James,
  personal communication.
Ogot and Kieran, 1969.
Osogo, 1966.
Roscoe, 1911.
Routledge, 1910, 1963.
Zarembka, David,
  personal communication.

SECTION 8

Chapter 25 PURE MATHEMATICS IN AFRICA

Al-Sa'di, Abd al-Rahmān ibn Abdullah:
  "A Timbuktu Scholar in Kano" in Hodgkin, 1960.
Barth, 1857, 1965.
Bernal, 1954.
Bivar and Hiskett, 1962.
Bivar, A. D. H.,
  personal communication about Muslim scholarship in West Africa.
Césaire, Aimé,
  "Cahier d'un Retour au Pays Natal". Présence Africaine, Paris, 1956.
Clapperton, 1829, 1966.
Davidson, 1960.
Diop, Cheikh Anta,
  "Negro Nations and Culture," in Collins.
Du Bois, 1965.
Eves, 1969.
Fullerton, Garry,
  "Education for Tomorrow" in Nolen, 1967.

Gay and Cole, 1967.
Herskovits, 1962.
Hiskett, 1967.
Hodgkin, 1960.
Hogben and Kirk-Greene, 1966.
Kennedy, E. S.,
  personal communication about Muslim African mathematicians.
Kenyatta, 1938, 1953.
Mchunguzi, 1969.
Mundy-Castle, A. C.,
  "Psychology" in Brokensha and Crowder, 1967.
Shaw, Thurstan,
  "Africa in Prehistory: Leader or Laggard?" *Jrl. of African History*, XII # 1 (1971),
  pp. 143–153. Shaw discusses the effects of ecological pressure and cultural
  isolation on Africa's cultural and technical development.
Shinnie, 1965.
Smith, H., 1952.
Strouhal, Eugen,
  "Evidence of the Early Penetration of Negroes into Prehistoric Egypt" *Jrl.
  of African History*, XII # 1 (1971), pp. 1–9.
Struik, 1967.
Struik, Dirk J.,
  personal communication on Muslim mathematicians.
White, 1956.

NOTE on Arabic books in Africa.

In 1854 Heinrich Barth wrote in his diary about a friend in Timbuktu (Volume III page 372.): "This [Arabic version] of Hippocrates had been a present from Captain Clapperton to Sultan Bello of Sokoto, from whom my friend had received it, among other articles, as an acknowledgment of his learning. I may assert, with full confidence, that those few books taken by the gallant Scotch captain into Central Africa have had a greater effect in reconciling the men of authority in Africa to the character of Europeans than the most costly present ever made to them."

BIOGRAPHICAL NOTES

Dr. Marianne Schmidl

Author of "Zahl und Zählen in Afrika" ("Number and Counting in Africa"),
Dr. Schmidl lived and worked in Vienna. All the information I have been
able to gather about her life and its tragic end is contained in the following
letter from Dr. Paul Spindler, of the Anthropological Society of Vienna:

> Dr. Marianne Schmidl was a student of Professor R. Pöch, the first
> Chairman of Anthropology in Vienna. She published two other very
> interesting works: "Foundations of the Cultures of the Nilotes," *Pro-
> ceedings of the Anthropological Society in Vienna*, Vol. 65, 1935, pages
> 86–127, and "Ancient Egyptian Techniques in African Spirally Woven
> Baskets," Essays for Pater Schmidt, *Anthropos*, Vienna, 1928, pages 645
> –654. I learned from Professor Hirschberg, who knew her personally,
> that Dr. Schmidl was distinguished for the collection of large amounts
> of material, and for precise documentation, and might have published
> many more works if she had not subjected her findings to such critical
> analysis.
>
> Her personal fate was a tragic one, due to political circumstances.
> In the midst of her professional career as librarian at the National
> Library in Vienna, she was persecuted because of her race, and evi-
> dently she was put to death. Dr. Hirschberg paid tribute to Dr. Schmidl
> in his lecture in celebration of the one hundredth anniversary of the
> Society [in 1970].

Dr. D. W. Crowe

Author of *Geometric Symmetries in African Art* (Part II of Chapter 14),
Dr. D.W. Crowe is Professor of Mathematics at the University of Wisconsin,
specializing in geometry. For several years he taught at the University of
Ibadan, Nigeria, and he was also a member of the Committee for the
Entebbe Program, a modern mathematics curriculum for the African
nations.

Dr. Crowe's article, "The Geometry of African Art. I. Bakuba Art,"
published in the *Journal of Geometry*,* includes drawings of one of each of
the nineteen patterns (seven strip, twelve plane) which actually appear in
the art of the Bakuba people. Similar complete information about Benin
art will soon be ready for publication. At present, he is gathering material
on the art of Cameroon, and on dyed cloths from various sources.

---

* *Journal of Geometry*, Basel. Vol. 1, No. 2, pages 169–182.

Frank E. Chapman, Jr.

Author of a collection of essays entitled *Science and Africa*, Frank E. Chapman, Jr. is a prisoner in a state penitentiary. Undaunted by failure to have *Science and Africa* published, Mr. Chapman has continued to work on the subsequent volumes, *Science, Society and Truth*, and *Race and Society*. We can only hope that such genius will soon receive the recognition it deserves, and that Mr. Chapman will be able to enjoy the relative freedom of life outside of prison.

The following excerpts from Chapman's autobiography throw some light on the misery of his existence and his determination to devote his life to "the search for truth, for this is my only true religion."

On October 5, 1961, I was sentenced to life imprisonment and fifty years for murder and robbery. I was nineteen years of age at this time. The circumstances which led to my present imprisonment greeted me the day I was born, and are shaping the destinies of millions of black men and women even now. These are the circumstances of material and spiritual poverty....

Presently, I am in prison, confined in a box-like cell, which is about six feet in width, seven and a half feet in length, and ten feet high.... Prison is a place where a few men acquire new dreams and new hopes, and where most men lose even the ability to hope for anything except a successful crime career.... But I left the courtroom with this conviction; I must know the truth about myself, I must become conscious of whatever forces have shaped my life and made me what I am. On that day life for me took a new and sudden turn. I wanted to live only to know myself better, and the world. I wanted to discover for myself my own humanity.

In many respects prison is almost an ideal place for study, for one has a great deal of leisure time.... First I enrolled in school, in the eighth grade. In a few months I graduated. This was the second time I had graduated from grade school. As soon as I thoroughly refreshed myself on the fundamentals, I began to study seriously. The first subject I attacked was mathematics. In a couple of months I was doing advanced algebra. I read and studied such books as Eric T. Bell's *Development of Mathematics*, Lancelot Hogben's *Mathematics for the Million*, and *Principia Mathamatica* by Bertrand Russell and Alfred N. Whitehead. After reading *Mathematics for the Million*, I had little difficulty in understanding the other books.

The study of mathematics led Chapman to natural science, then to philosophy and African culture and history.

I was determined to know everything the human race had done, as well as everything it is doing.

In 1964 I decided to write a book demonstrating the scientific contributions of African peoples to mankind. In June 1964 I wrote a brief synopsis of my ideas, and submitted it in the form of an article entitled "Mathematics in Antiquity" to the editors of *Freedomways* Magazine in New York. . . . Mr. John H. Clarke, an editor of *Freedomways*, got the article accepted by a Parisian magazine entitled *Présence Africaine*. After my article was accepted, I began to work on my book. . . .

Within a year *Science and Africa* was finished. Upon finishing I wrote the following: I would be the first to admit that this is a stupendous task for anyone to undertake, but it is a little more than that for a man in prison endeavoring to write and carry on scientific research under severe and trying circumstances; yet, I was and am determined to get the job done, and I am confident that this work (*Science and Africa*) will provide a useful and socially constructive outlet for the outpouring of that social energy I have inherited. I have spared no effort in the endeavor to make myself fully cognizant of the impact the modern world has had on me and my fellow human beings. And out of this has been born an outlook which has enabled me to link up my own individual social experience of mankind. . . . Such an outlook is not merely born out of dreams and fantastic longings for a better way of life; it is born out of man's efforts to grapple with his objective existence and it is the culmination of all his constructive achievements. . . .

Reflections of this nature have led me to the conclusion that the most pressing task of African philosophers, scientists, and political leaders is the laying of ideological foundations by assessing the nature of Africa's role in the modern world, and thereby arriving at some concrete conception of her historical mission. . . . My *Science and Africa* deals with only one of the many significant effects of European imperialism. After conquering the world, the white man has falsely claimed that he and he alone is the benefactor of human civilization. In this book I try to destroy, theoretically, the white man's racial monopoly in the field of natural science; and to the extent that I succeed I will have contributed to the happiness and well-being of mankind.

I sent the manuscript to John Clarke, editor of *Freedomways* in the summer of 1966. *Freedomways* published some excerpts from *Science and Africa*. . . .

Through knowledge and understanding, I have freed myself from my past errors and have created for myself a bright and beautiful future. My great ambition in life is to try to keep others from repeating the mistakes I have made and to free my brothers and sisters from the humiliation of being black and poor. . . . I cannot accept the argument that I and I alone am responsible for what happened, for the crimes I committed. If that be the case, then the major premise of modern sociology and psychiatry is false. However, what I wish to bring out is the fact that I am a new man, with new interests, dreams and ambitions.

. . . I am not trying to conceal anything I have done behind glittering generalizations; I only want the reader of these cold and lifeless words to understand that ours is not the story of just a solitary man who presently destroyed his chances to be a fine fellow, but a man born into a social situation that chokes and cripples man's infinite capacity to produce and create. Therefore, what I have done, and what has been done to me, is in no way purely unique.

To tell the truth, my story is the story of millions. One black brother has appropriately called us "The Wretched of the Earth."

BIBLIOGRAPHY

Abraham, Roy. *The Principles of Ibo*. Ibadan: U. of Ibadan, 1967.

Achebe, Chinua. *Arrow of God*. New York: John Day, 1967.

——— . *Things Fall Apart*. New York: Astor-Honor, 1959.

Adamson, Joy. *Born Free*. New York: Hillman, McFadden, 1961.

Ajayi, J.F. Ade and Esprie, Ian (eds.). *A Thousand Years of West African History*. Ibadan: Ibadan U. Press, 2nd ed., 1968.

Almeida, Antonio de. "Sobre a Mathemática dos Indígenos da Guiné Portuguesa." *Boletim Cultural da Guiné Portuguesa* (Lisbon), No. 6 (April 1947).

Andrews, U.S. *Magic Squares and Cubes*. New York: Dover, 2nd ed., 1960.

Armstrong, R.G. *Comparative Wordlist of Five Igbo Dialects*. Occasional Publication No. 5, Inst. of African Studies, U. of Ibadan, 1967.

——— . *Yoruba Numerals*. Ibadan: Oxford U. Press, 1962.

Armstrong, R.G., Olayemi, V., and Adu, B. "Ekiti Traditional Dirge of Lt. Col. Adekunle Fajuyi's Funeral." *African Notes* (Ibadan) 5:2 (January 1969), pp. 63ff.

Atkins, W.J. "Notes on the Concords and Classes of Bantu Numerals." *African Language Studies* (London) 2 (1961), pp. 42–48.

Babayemi, S.O. "Oyo Ruins." *African Notes* (Ibadan) 5:1 (1968), pp. 8–11.

Ball, W.W. Rouse. *Mathematical Recreations and Essays*. London: Macmillan, 1928.

Barth, Heinrich. *Travels and Discoveries in North and Central Africa*, 1857. Reprinted, London: Frank Cass, 1965.

Bascom, William. *Ifa Divination*. Bloomington: Indiana U. Press, 1969.

Basden, G.T. *Among The Ibos of Nigeria*. New York: Barnes & Noble, 1921, 1966.

——— . *Niger Ibos*. London: Frank Cass, 1938, 1966.

Bastin, Marie-Louise. *Art Decoratif Tshokwe*. Lisbon: Museu do Dundo, 1961.

Bell, Robert C. *Board and Table Games*, Vol. I. New York: Oxford U. Press, 1960.

ben-Jochannan, Yosef. *Black Man of the Nile*. New York: Alkebu-lan, 1970.

Bernal, J.D. *Science in History*. 2 Vols. New York: Cameron, 1954.

Bivar, A.D.H. and Hiskett, M. "Arabic Literature of Northern Nigeria Up to 1804." *Bulletin of SOAS* (London U.) XXV:1 (1962), pp. 104–148.

Blacking, John. *Venda Children's Songs*. Johannesburg: Witwatersrand U. Press, 1967.

Bleeker, Sonia. *The Ibo of Biafra*. New York: Morrow, 1969.

——— . *The Masai, Herders of East Africa*. London: Dennis Dobson, 1964.

Bohannan, Paul. *Africa and Africans*. Garden City, N.Y.: The Natural History Press, 1964.

Bohannan, Paul and Bohannan, Laura. *Tiv Economy*. London: Longmans, 1968.

Bosman, William. *A New and Accurate Description of the Coast of Guinea*, 1704. New York: Barnes & Noble, new ed. 1967.

Boyer, Carl B. *History of Mathematics*. New York: Wiley, 1968.

Brain, James L. *Basic Structure of Swahili*. Syracuse, N.Y.: Syracuse U. (undated).

Brokensha, David and Crowder, Michael (eds.). *Africa in the Wider World*. Oxford: Pergamon Press, 1967.

Cajori, Florian. *History of Elementary Mathematics.* New York: Macmillan, 1896, 1917.
—— . *A History of Mathematics.* New York: Macmillan, 1929.
Chapman, Frank E., Jr. *Science and Africa* (unpublished manuscript). Excerpts appeared in *Freedomways* 6:3 (Summer 1966), pp. 235-245.
Clapperton, Hugh. *Journal of a Second Expedition into the Interior of Africa,* 1829. Reprinted, London: Frank Cass, 1966.
Clegg, Legrand H., II. "The Beginning of the African Diaspora: Black Men in Ancient and Medieval America?" *Current Bibliography on African Affairs* (Washington, D.C.) 2:12 (December 1969) , pp. 13-34 .
Collins, Robert O. *Problems in African History.* Englewood Cliffs, N.J.: Prentice-Hall, 1968.
Conant, Levi L. *The Number Concept.* New York: Macmillan, 1896.
Crowder, Michael. *The Story of Nigeria.* London: Faber & Faber, 1966.
—— . *West Africa Under Colonial Rule.* Evanston, Ill.: Northwestern U. Press, 1968.
Crowe, Donald W. "The Geometry of African Art I. Bakuba Art." *Journal of Geometry* (Birkhäuser Verlag, Basel) 1:2 (1971) , pp. 169–183.
Culin, Stewart. "Mancala, The National Game of Africa." Washington, D.C.: Smithsonian Institution, June 30, 1894, pp. 597–606.

Dantzig, Tobias. *Number, The Language of Science.* London: George Allen & Unwin, 1940.
Davidson, Basil. *African Genius.* Boston: Atlantic, Little Brown, 1969.
—— . *African Kingdoms.* New York: Time-Life, 1966.
—— . *The African Past.* Boston: Little Brown, 1964.
—— . *The African Slave Trade.* Boston: Little Brown, 1961.
DeGraft-Johnson, J.C. *African Glory.* New York: Walker, 1954.
De Heinzelin, Jean. "Ishango." *Scientific American 206:6 (June* 1962), pp. 105-114.
Delafosse, Maurice. *La Langue Mandingue et ses Dialectes,* Vol. I. Paris: Librairie Orientaliste, Paul Guenther, 1929.
—— . "La Numération chez les Nègres." *Africa* I:3 (1928), pp. 387-390.
Delano, Issac O. *Conversations in Yoruba and English.* New York: Praeger, 1963.
—— . *The Soul of Nigeria.* London: T. Werner Laurie, 1937.
Dennett, R.E. *Nigerian Studies.* London: Frank Cass, 1910, 1968.
Dike, K. Onwuka. *Trade and Politics in the Niger Delta.* Oxford: Clarendon Press, 1956.
Dohne, Rev. J.L. *Zulu-Kafir Dictionary.* Capetown: C.F.M., 1857.
Driberg, J.H. "The Game of Choro." *Man* (London) 27 (September 1927).
Du Bois, W.E.B. *The World and Africa.* New York: International Publishers, 1965.
Dundas, Charles. "Chagga Time Reckoning." *Man* (London) 26 (August 1926), pp.140-143.
Dyson-Hudson, N. *Karimojong Politics.* Oxford: Clarendon Press, 1966.

Egharevba, Jacob. *A Short History of Benin.* Ibadan: Ibadan U. Press, 1960.
Einzig, Paul. *Primitive Money.* Oxford: Pergamon Press, 2nd ed., 1966.
Eves, Howard. *An Introduction to the History of Mathematics.* New York: Holt, Rinehart & Winston, 2nd ed., 1969.
—— *In Mathematical Circles.* Boston: Prindle, Weber & Schmidt, 1969.

Fagg, William and Willett, Frank. "Ancient Ife." *Odu* (Ibadan), No. 8 (1960) , pp. 21-35.
Farquhar, J.H. "The African Hand." *Nada* (Salisbury), No. 25 (1948), pp. 25-28; No. 27 (1950), pp. 25-30.

Fitch, J.M. and Branch, D.P. "Primitive Architecture and Climate." *Scientific American* 203:6 (December 1960), pp. 134–144.

Forde, Daryll. *The Yoruba-Speaking Peoples of South Western Nigeria.* London: International African Inst., 1951.

Frobenius, Leo. *Voice of Africa,* Vol. I. New York: Benjamin Blom, 1913, 1968.

Fuja, Abayemi. *1400 Cowries (Yoruba Tales).* London: Oxford U. Press, 1962.

Gay, John and Cole, Michael. *The New Mathematics and an Old Culture, A Study of Learning among the Kpelle of Liberia.* New York: Holt, Rinehart & Winston, 1967.

Gecaga, B. Mareka and Kirkaldy-Willis, W.H. *A Short Kikuyu Grammar.* London: Macmillan, 1953.

Greenberg, J.H. *The Languages of Africa.* Bloomington: Indiana U. Press, 2nd ed., 1966.

Griaule, Marcel. *Conversations with Ogotemmêli.* London: Oxford U. Press, 1965.

Gulliver, Philip H. "Counting with the Fingers by Two East African Tribes." *Tanganyika Notes,* No. 51 (December 1958), pp. 259–262.

——— . *Social Control in an African Society.* Boston: Boston U. Press, 1963.

Gwarzo, Hassan Ibrahim. "The Theory of Chronograms as expounded by the 18th century Katsina Astronomer-Mathematician Muhammad B. Muhammad." *Research Bulletin of the Center of Arabic Documentation* (U. of Ibadan) 3:2 (July 1967), pp. 116–123.

Hall, Barbara (ed.). *Tell Me Josephine.* New York: Simon & Schuster, 1964.

Hall, R. de Z. "Bao." *Tanganyika Notes,* No. 34 (January 1953), pp. 57–61

Hattersley, C.W. *The Baganda at Home.* London: Frank Cass, 1908, 1968.

Herskovits, M.J. *Dahomey, an Ancient West African Kingdom.* Evanston, Ill.: Northwestern U. Press, 1967.

——— . *Economic Anthropology.* New York: Knopf, 1952.

——— . *The Human Factor in Changing Africa.* New York: Knopf, 1962.

——— . "The Numerical System of the Kru." *Man* (London) 39, Item 148 (1939), pp.154-55.

——— . "Wari in the New World." *Journal of the Royal Anthropological Institute* 62 (1932), pp. 23-37.

Hill, Polly. "Notes on the Traditional Market Authority and Market Periodicity in West Africa." *Journal of African History* VII:2 (1966), pp. 295–311.

Hintze, Fritz. "New Researches in Old African Culture." *Scientific World* (World Federation of Scientific Workers, New York), No. 1 (1965), pp. 16–20, 30.

Hiskett, M. "The Arab Star-Calendar and Planetary System in Hausa Verse." *Bulletin of SOAS* (London U.) XXX (1967), pp. 158-176.

Hodgkin, Thomas. *Nigerian Perspectives.* London: Oxford U. Press, 1960.

Hogben, L. *Mathematics in the Making.* London: Rathbone, 1960.

Hogben, S. and Kirk-Greene, A.H.M. *The Emirates of Northern Nigeria.* London: Oxford U. Press, 1966.

Hollis, A.C. *The Masai.* Oxford: Clarendon Press, 1905.

——— . *The Nandi.* Oxford: Clarendon Press, 1909.

Howeidy, A. *A Concise Hausa Grammar.* London: George Ronald, 2nd ed., 1959.

Hurel, Eugene. *Grammaire Kinyarawanda.* Ruanda: Kabgayi, 1951.

Ibn Battuta, Muhammad. *Travels in Asia and Africa,* 1325-1354. Trans. and Ed., H.A.R. Gibbs. New York: Robert M. McBridge, 1929.

Ibn Muhammad, Muhammad. *Bahjat al-āfāq*, 1732. Portions of manuscript in the SOAS, U. of London.

Idowu, E. Bolaji. *Olodumare, God in Yoruba Belief.* London: Longmans, 1962.

Igwe, G.E. and Green, M.M. *Igbo Language Course.* Ibadan: Oxford U. Press, 1967.

Innes, Gordon. *A Practical Introduction to Mende.* London: Luzac, 1967.

Jackson, John G. *Introduction to African Civilization.* New York: University Books, 1970.

Jacottet, E. "Grammar of the Sesuto Language." *Bantu Studies* III (January 1927).

James, George G.M. *Stolen Legacy.* New York: Philosophical Library, 1954.

Jeffreys, M.D.W. "The Cowry Shell." *Nigeria*, No. 15 (1938), pp. 221–226.

Jensen, Ad E. *Im Lande des Gada.* Stuttgart: Verlag Strecker u. Schröder, 1936.

Johnson, Marion. "The Cowrie Currency of West Africa" (2 parts). *Journal of African History* XI:1 (1970); pp. 17-49; XI:3 (1970), pp. 331-353   70), pp. 331–353

Johnson, Samuel. *The History of the Yorubas.* London: Routledge and Kegan Paul, 1921, 1966.

Josephy, Alvin M., Jr. (ed.). *Horizon History of Africa.* New York: American Heritage, 1971.

Kenyatta, Jomo. *Facing Mount Kenya.* London: Secker & Warburg, 1938, 1953.

Kirk-Greene, A.H.M. "The Major Currencies in Nigerian History." *Journal of the Historical Society of Nigeria* II:1 (December 1960), pp. 132–150

Kramer, Edna. *The Main Stream of Mathematics.* New York: Oxford U. Press, 1951.

Lagercrantz, Sture. "African Tally Strings." *Anthropos* (Institutum Anthropos, Germany), No. 63 (1968), pp. 115-128.

Lanning, E.C. "Rock-cut Mweso Boards." *Uganda Journal* 20:1 (March 1956), pp. 95–98.

Lasebikan, E.L. "Brazilians Adopt Yoruba Game." *West Africa*, No. 2391 (March 30, 1963).

Latham, A.J.H. "Currency, Credit and Capitalism on the Cross River in the Pre-Colonial Era." *Journal of African History* XII:4 (1971), pp. 599-604.

Leiris, Michel and Delange, Jacqueline. *African Art.* New York: Golden Press, 1968.

Leuzinger, Elsy. *Africa, the Art of the Negro Peoples.* New York: Crown Publishers, 1960.

Lévy-Bruhl, Lucien. *How Natives Think.* New York: Washington Square Press, 1966. (Original French ed., 1910.)

Lindblom, Gerhard. *The Akamba.* Uppsala: J.A. Lundell, 1920.

Lloyd, P.C., Awe, B. and Mabogunje, A.L. *The City of Ibadan.* London and New York: Cambridge U. Press, 1967.

Luschan, Felix von. *Die Alterthümer von Benin.* Berlin: Hecker, 1968.

Mann, A. "On the Numeral System of the Yoruba Nation." *Journal of the Royal Anthropological Society* XVI (1887), pp. 59-64.

Marshack, Alexander. *The Roots of Civilization.* New York: McGraw-Hill, 1972.

Mathews, H.F. "Duodecimal Numeration in Northern Nigeria." *The Nigerian Field* XXIX:4 (October 1964), pp. 181–191.

Matthews, J.B. "Notes on Some African Stone Games." *Nada* (Salisbury) IX:1 (1964), pp. 64-68.

Mbiti, John S. *African Religions and Philosophy.* New York: Praeger; London: Heinemann, 1969.

——— . *English-Kamba Vocabulary.* Kampala: East Africa Literature Bureau, 1959.

McCall, Daniel F. *Africa in Time Perspective.* New York: Oxford U. Press, 1968.

Mchunguzi, Salim. "Kenya's Political Evolution," in *Inside Kenya Today.* Kenya Dept. of Information, March 1969 , pp. 24–27.

Menniger, Karl. *Number Words and Number Symbols.* Cambridge, Mass.: M.I.T. Press, 1969. (English translation of *Zahlwort und Ziffer,* Göttingen: Vandenhoeck u. Ruprecht, 1958.)

Migeod, W.L. *Languages of West Africa.* 2 Vols. London: Kegan, Paul, Trench, Trubner, 1911.

Morton-Williams, Peter. "An Outline of the Cosmology and Cult Organization of the Oyo Yoruba." *Africa,* No. 34 (1964) , pp. 243–261.

Mott-Smith, G. *Mathematical Puzzles.* New York: Dover, 2nd ed., 1954.

Munro, David A. *English-Edo Word List.* Occasional Publication No. 7, Inst. of African Studies, U. of Ibadan, 1967.

Murdock, G.P. *Africa: Its People and Their Culture History.* New York: McGraw-Hill, 1959.

Murray, H.J.R. *A History of Board Games Other than Chess.* Oxford: Clarendon Press, 1952.

Neugebauer, O. *The Exact Sciences in Antiquity.* New York: Harper, 1962.

Ngugi, James. *The River Between.* London: Heinemann, 1965.

Nilsson, M.P. *Primitive Time Reckoning.* Lund: C.W.K. Gleerup, 1920.

Nolen, Barbara. *Africa is People.* New York: Dutton, 1967.

Nsimbi, M.B. *Omweso, A Game People Play in Uganda.* Occasional Paper No. 6, African Studies Center, U. of California at Los Angeles, 1968.

A.V.O. "The Hima Method of Counting." *Uganda Journal* 4:1 (July 1936), p. 91.

Ogot, B.A. *History of the Southern Luo.* Nairobi: East Africa Publishing House, 1967.

Ogot, B.A. and Kieran, J.A. *Zamani.* Nairobi: East Africa Publishing House, 1968.

Ojike, Mbonu. *My Africa.* New York: John Day, 1946.

Ojo, G.J. Afolabi. *Yoruba Culture.* London: U. of Ife and U. of London Press, 1966.

Olderogge, D.A. "Ancient Scripts from the Heart of Africa." *The UNESCO Courier* 19:3 (March 1966), pp. 25–29.

Osogo, John. *A History of the Baluyia.* Nairobi: Oxford U. Press, 1966.

Pankhurst, Richard. "A Preliminary History of Ethiopian Measures, Weights and Values." *Journal of Ethiopian Studies* III: 1 & 2 (January & July 1969).

Parrinder, Geoffrey. *African Mythology.* Feltham, England: Paul Hamlyn, 1967.

Phillipson, D.W. "Notes on the Later Prehistoric Radiocarbon Chronology of Eastern and Southern Africa." *Journal of African History* XI:1 (1970), pp. 1–15.

Prussin, Labelle. *Architecture in Northern Ghana.* Berkeley and Los Angeles: U. of California Press, 1969.

Rattray, R.S. *The Ashanti,* Vol. 1. London: Oxford U. Press, 1923, 1969.

——— . *Religion and Art in Ashanti,* Vol. 2. London: Oxford U. Press, 1927, 1954.

Raum, O.F. *Arithmetic in Africa.* London: Evans Bros., 1938.

——— . "The African Chapter in the History of Writing." *African Studies* (Johannesburg) 2:4 (December 1963), pp. 179–192.

Ritchie-Calder, Lord. "Conversion to the Metric System." *Scientific American* 223:1 (July 1970) , pp. 17–25.

Rohrbough, Lynn. *Ancient Games*, 1936. *Count and Capture*, 1955. Delaware, Ohio: Co-operative Recreation Service.

Roscoe, John. *The Baganda.* London: Frank Cass, 1911, 1965.

——. *The Northern Bantu.* New York: Barnes & Noble, 1915, 1966.

Routledge, W.S. and Routledge, K. *With a Prehistoric People, The Akikuyu of British East Africa.* London: Frank Cass, 1910, 1968.

Ryder, A.F.C. "Dutch Trade on the Nigerian Coast during the Seventeenth Century." *Journal of the Historical Society of Nigeria* III:2 (December 1965) , pp. 195–210.

Sawyer, W.W. "The Game of Oware." *Scripta Mathematica* (Yeshiva U.) XV:2 (1949).

Schmidl, Marianne. "Zahl und Zählen in Afrika." *Mitteilungen der Anthropologischen Gesellschaft in Wien* 45 (1915), pp. 165-209.

Seidenberg, A. "Ritual Origin of Counting," in *Archive for the History of Exact Sciences.* Berlin: Springer Verlag, 1962, pp. 1-40.

——. "The Diffusion of Counting Practices." *University of California Publications in Mathematics* III:4 (1960) , pp. 215–299.

Shaw, Thurstan. "Radiocarbon Dates from Nigeria." *Journal of the Historical Society of Nigeria* III:4 (June 1967), pp. 743–751.

Shinnie, Margaret. *Ancient African Kingdoms.* London: Edward Arnold, 1965.

Siegel, Morris. "A Study of West African Carved Gambling Chips." *Supplement to American Anthropologist* 42:4, Part 2 (1940).

Smeltzer, Donald. *Man and Number.* New York: Emerson Books Inc., 1958.

Smith, David Eugene. *History of Mathematics*, Vol. 1. New York: Dover, 1958. (First ed., 1923.)

Smith, Homer W. *Man and His Gods.* New York: Grossett & Dunlap, 1952.

Stevens, Phillips. "Orisha-nla Festival." *Nigeria*, No. 90 (September 1966) , pp. 184–199.

Struik, Dirk J. *A Concise History of Mathematics*, Vol. 1. New York: Dover, 1967.

——. "On Ancient Chinese Mathematics." *The Mathematics Teacher* (National Council of Teachers of Mathematics) 56:6 (October 1963), pp. 424–432.

Thomas, N.W. *Anthropological Report on the Edo-Speaking Peoples of Nigeria.* London: Harrison & Sons, 1910.

——. "Duodecimal Base of Enumeration." *Man* (London) 20 (1920), pp. 25-29.

Torday, Emil. *On the Trail of the Bushongo.* Philadelphia: Lippincott, 1925.

Torrey, Volta. "Old African Numbers Game?" in *Science Digest.* New York: Hearst, 1963.

Trowell, Margaret. *African Design.* London: Faber & Faber, 1960.

Tucker, A.N. and Mpaayei, J. Tempo ole. *A Maasai Grammar.* London: Longmans, Green, 1955.

Turnbull, Colin. *The Forest People.* New York: Simon & Schuster, 1961.

Turner, Lorenzo D. *Africanisms in the Gullah Dialect.* New York: Arno Press, 1949, 1969.

Ukwu, Ukwu I. "The Development of Trade and Marketing in Iboland." *Journal of the Historical Society of Nigeria* III:4 (June 1967) , pp. 647–662.

Van der Waerden, B.L.   *Science Awakening.* New York: Oxford U. Press, 1961.
Van Sicard, Harold.   "The Rhodesian Tally." *Nada* (Salisbury), No. 31 (1954), pp. 52-54.

Wayland, E.J.   "Notes on the Board Game Known as Mweso in Uganda." *Uganda Journal* 4:1 (July 1936) , pp. 84–89
White, Leslie A.   "The Locus of Mathematical Reality: An Anthropological Footnote," in *The World of Mathematics,* ed. J. Newman, Vol. 4. New York: Simon & Schuster, 1956.
Wilder, R.L.   *Evolution of Mathematical Concepts.* New York: Wiley, 1968.
Willett, Frank.   *African Art.* London: Thames & Hudson; New York; Praeger, 1971.
Williams, Chancellor.   *The Destruction of Black Civilization: Great Issues of Race from 4500 B.C. to 2000 A.D.* Dubuque, Iowa: Kendall/Hunt Publishing Co., 1971.
Williamson, John.   "Dabida Numerals." *African Studies* (Johannesburg), No. 4 (1943), pp. 215-6.
Williamson, Kay and Timitimi, A.O.   "A Note on Ijo Number Symbolism." *African Notes* 5:3 (January 1970) , pp. 9–16

Ziervogel, D., Louw, J.A. and Ngidi, J.   *Handbook of the Zulu Language.* Pretoria: Van Schaik, 1967.

The index has been divided into two sections:

Section A is based upon peoples, languages, and geographical divisions in Africa.

Section B is based upon subject, and includes a selected number of authors whose works are quoted in the text. A complete list of authors may be found in the chapter notes and bibliography.

Section A. Africa: Peoples, Languages, Geography
The name of each ethnic group is followed by the name of the country where it is found (in boldface). However, since geographical lines were generally drawn by the European colonial powers without regard to the peoples involved, many groups are found in more than one country.

Section B. Subjects and Selected Authors

The reader may refer to the chapter notes and bibliography for a complete list of authors.